Foreword

"Oxalate Metabolism in Relation to Urinary Stone" is the third monograph to appear in the "Bloomsbury Series". Edited by Alan Rose, the book describes the current clinical and biochemical features of oxalate metabolism. Its content and direction fulfil the goals of the Series emphasising the strong links between basic science and clinical medicine.

London   Jack Tinker

30 March 1988

OXALATE METABOLISM IN RELATION TO URINARY STONE

G. ALAN ROSE (Ed.)

With 98 figures

Springer-Verlag
London Berlin Heidelberg New York
Paris Tokyo

G. Alan Rose, MA, MD, FRCP, FRCPath, FRSC
Consultant Chemical Pathologist, St. Peter's Hospitals and
Institute of Urology, London, St. Paul's Hospital, 24, Endell
Street, London WC2H 9AE

Series Editor
Jack Tinker, BSc, FRCS, FRCP, DIC
Director, Intensive Therapy Unit, The Middlesex Hospital,
London W1N 8AA UK

Cover illustration: Calcium phosphate crystals appearing as
fibrils.

ISBN-13:978-1-4471-1628-8   e-ISBN-13:978-1-4471-1626-4
DOI: 10.1007/978-1-4471-1626-4

British Library Cataloguing in Publication Data
Rose, G. Alan (George Alan)
Oxylate metabolism in relation to urinary stone.—(The Bloomsbury series in
clinical science).
1. Man. Urinary tract. Calculi   I. Title   II. Series 616.6'22

Library of Congress Cataloging-in-Publication Data
Oxalate metabolism in relation to urinary stone/G. Alan Rose (ed.). p. cm.—
(The Bloomsbury series in clinical science)
Edited proceedings of a workshop held in 1987.
Includes bibliographies and index.
ISBN-13:978-1-4471-1628-8
1. Calculi, Urinary—Congresses.   2. Oxalic acid—Metabolism—Disorders
—Congresses.   3. Calcium oxalate—Metabolism—Disorders—Congresses.
I. Rose, G. Alan.   II. Series. [DNLM: 1. Oxalates—metabolism—congresses.
2. Urinary Calculi—metabolism—congresses.   WJ 100 098 1987] RC916.093
1988   616.6'22—dc19

© Springer-Verlag Berlin Heidelberg 1988
Softcover reprint of the hardcover 1st edition 1988

Filmset by Photo·Graphics, Honiton, Devon

2128/3916–543210

Preface

The first oxalate workshop was held in London in 1979 and the proceedings published privately by the Wellcome Foundation. At that time the importance of urinary oxalate as a risk factor more important for calcium oxalate stone formation than urinary calcium had been recognized. Nevertheless measurements of urinary oxalate still left a lot to be desired and in particular the non-enzymatic conversion of ascorbate to oxalate had not been rediscovered so that many measurements must have been wrong. Plasma oxalate was still difficult or impossible to measure by any reasonable, accessible methods and consequently there was still much argument and speculation about the handling of oxalate by the kidneys. A lot of work has been performed in the last eight years on oxalate metabolism and it therefore seemed to the organisers to be a good time to hold a second oxalate workshop. The first oxalate workshop was an off-shoot of the Bonn–Vienna Symposia on Urolithiasis and it consisted of a series of papers submitted by British and Europeans working in the field. This second workshop, however, consisted of a number of review papers given by authors specially selected for their expertise in particular aspects of oxalate research. An attempt has been made to cover both clinical and laboratory aspects and it is therefore hoped that this volume will appeal to both doctors and biochemists.

A workshop such as this can only reach a limited audience and its value is greatly enhanced by publication which gives access to a far greater audience. I am therefore grateful to Springer-Verlag for having agreed to publish the edited proceedings as this volume. I would also like to acknowledge financial assistance from the Sigma Chemical Company.

Department of Pathology,   G. A. Rose
St. Paul's Hospital, London WC2H 9AE

December 1987

Contents

Contributors

T. M. Barratt, FRCP
Institute of Child Health, Department of Paediatric Nephrology,
Institute of Child Health, 30 Guildford Street, London WC1N
1EH

M. G. Dillon, MB, FRCP, DCH
Consultant Paediatric Nephrologist, Renal Unit, Hospital for Sick
Children, Great Ormond Street, London WC1N 3JH

P. C. Hallson, BSc, PhD, MRSC
Institute of Urology and St Peter's Hospitals, Department of
Biochemistry, St Paul's Hospital, 24 Endell Street, London WC2H
9AE

G. P. Kasidas, MSc, PhD
Department of Biochemistry, St Paul's Hospital, 24 Endell Street,
London WC2H 9AE

P. K. S. Liu, BSc, PhD
Department of Biochemistry, St Paul's Hospital, 24 Endell Street,
London WC2 9AE

M. A. Mansell, MD, MRCP
Consultant Nephrologist, St Peter's Hospitals and Institute of
Urology, St Philip's Hospital, Sheffield Street, London WC2 2EX

D. S. Rampton, DPhil, MRCP
Consultant Physician, Newham General and The London Hospital,
Newham General Hospital, Glen Road, London E13 8SL

G. A. Rose, MA, DM, FRCP, FRCPath, FRSC
Institute of Urology and St Peter's Hospitals, St Paul's Hospital,
24 Endell Street, London WC2H 9AE

C. T. Samuell, MIBiol, MCB
Institute of Urology and St Peter's Hospitals, Department of
Biochemistry, St Paul's Hospital, 24 Endell Street, London WC2H
9AE

M. Sarner, MD, FRCP
Consultant Physician, University College Hospital, Gower Street,
London WC1E 6JJ

R. S. Trompeter, MB, MRCP
Consultant Paediatric Nephrologist. Royal Free Hospital London
and Hospital for Sick Children, Great Ormond Street, London
WC1N 3JH

Vanessa von Sperling, BSc
Clinical Fellow, Department of Paediatric Nephrology, Institute
of Child Health, 30 Guildford Street, London WC1N 1EH

R. E. W. Watts, MD, DSc, PhD, FRCP, FRSC
Head, Division of Inherited Metabolic Diseases, MRC Clinical
Research Centre, Watford Road, Harrow, HA1 3UJ

Chapter 1

Introduction

G. A. Rose

Oxalate is an interesting chemical from the biological point of view. In the plant world it has important roles:

1. To precipitate as calcium oxalate excess calcium absorbed by the root systems of plants or fungi
2. To provide a form of exoskeleton by its extrusion from the plant or fungus in the form of calcium oxalate
3. To make the plant less acceptable to animals as fodder by virtue of the taste of free oxalic acid

In man, on the other hand, oxalate confers no advantages but some distinct disadvantages. Thus it cannot be metabolised within the body and any oxalate absorbed or generated is eliminated by the kidneys to appear in the urine where it may cause calcium oxalate crystalluria and/or stones. Hence the production or absorption of oxalate in man would appear to be disadvantageous and one might well wonder why they occur at all. Omniverous man is protected against the oxalate present in some of the plant life he eats in a number of ways. First, oxalate absorption from the gastrointestinal tract is normally rather an inefficient process (Marshall et al. 1972; Finch et al. 1981) which takes place by passive diffusion only, the process not being energy-dependent (Binder 1974); (although see Chap. 7). Furthermore, the gastrointestinal tract houses bacteria which can degrade some of the oxalate ingested (Allison et al. 1986). Nevertheless, significant quantities of oxalate are absorbed and make a substantial contribution to its urinary excretion level. It is widely stated in the literature that most of the urinary oxalate arises from endogenous production, mainly from glycine and ascorbate, a thesis that seems strange from a teleological standpoint in view of the distinct disadvantages arising from this production. In fact this widespread concept may be wrong. First, while on a low-oxalate diet most of the urinary oxalate may be endogenous in origin since the actual 24-h urinary oxalate values then are quite low (about 0.2 mmol/24h or less). Second, a lot of the early work showing conversion of ascorbate to oxalate may have been rendered invalid by the finding that ascorbate spontaneously and non-enzymatically converts to oxalate especially at alkaline pH (Herbert et al. 1973; Rose 1985). The conversion may therefore have taken place not in the body, but in the test tube.

The early chapters of this volume deal with methods for measuring oxalate and glycollate levels in urine and in plasma. A methods section is clearly of direct importance to the biochemist, but it is important for clinicians to realise that most hospital laboratories either do not measure oxalate at all or else do it rather badly as shown by the external quality assurance scheme described in Chapter 3. Almost no hospital laboratories measure glycollate, which means that the diagnosis of mild metabolic hyperoxaluria (Chap. 8) cannot be made. Clinicians need to encourage their biochemists to invest in satisfactory methods so that hyperoxaluria will be found and treated where it exists. Later chapters deal with the hyperoxaluric states and the ways whereby they can be treated and the benefits that can result from their treatment. It is hoped that biochemists reading these chapters will perceive the need for precise measurements of urinary glycollate and oxalate and then set up methods for making these measurements in their own laboratories.

Secondary hyperoxaluria comprises an important group of causes of urolithiasis and this is reviewed in Chap. 7. Most of these conditions are easily treated, but the hyperoxaluria of steatorrhoea can be remarkably resistant to treatment and therefore this has been considered in more depth. It has been thought for a long time that crystalluria must be linked in some way to the formation of urinary stone although the intricacies of the link still require a lot of study. This aspect is reviewed in some depth in Chapter 9 and many new and original observations using newly developed techniques are included, and fresh light is shed on the subject. The last chapter (Chap. 11) deals with pyridoxine metabolism and is included here because urinary oxalate levels may be lowered by pyridoxine administration in primary hyperoxaluria and in mild metabolic hyperoxaluria. However, not all the cases respond and it seems important to establish whether pyridoxine metabolism is different in the responders and the non-responders.

Ideas about renal transplantation for primary hyperoxaluria are in a particularly violent state of flux at the moment and are reviewed in Chap. 10. Only a few years ago renal transplantation was considered unhelpful as experience showed that the grafted kidney would quickly undergo oxalosis (Wilson 1975). More recently, with high doses of pyridoxine and vigorous dialysis therapy renal transplantation has become possible (O'Regan et al. 1980; Scheinman et al. 1984). Even more recently, liver transplantation has been proposed and carried out (Watts et al. 1987) on the grounds that the excessive oxalate production takes place in the liver (Danpure et al. 1987). If this is true and if the urinary clearance of oxalate remains approximately the same as clearance of creatinine both in renal failure and in primary hyperoxaluria (Kasidas and Rose 1986) then the plasma oxalate/creatinine ratio should be higher than normal in primary hyperoxaluria by a factor of three to five or more.

Plasma Oxalate/Creatinine Ratio

Table 1.1 summarises the results of measuring the plasma oxalate/creatinine ratio in normal subjects and in patients with chronic renal failure. It can be seen that in 99 samples from 10 normal subjects the overall mean value was 0.0224 (SD 0.0032) giving a normal range of 0.0160–0.0288. In 20 samples from patients with chronic renal failure (Kasidas and Rose 1986) the mean ratio was 0.025 which is not significantly different from that of the normal subjects. Section 1 of Table 1.1 shows the ratio from 15 patients with primary hyperoxaluria arranged in descending order (0.188–0.046) all of which are significantly raised above normal. The lowest values are amongst those patients showing some response to pyridoxine. As the plasma oxalate/creatinine ratio is independent of the degree of renal function it can be used as an indicator of whether a patient has primary hyperoxaluria. This is particularly useful when urine cannot be obtained and in children when the oxalate/creatinine ratio in urine is difficult to interpret (Kasidas and Rose 1987).

Renal Hyperoxaluria?

Section 3 of Table 1.1 shows results from two siblings who seem to fall in a different category. Both had quite severe primary hyperoxaluria as shown by raised levels of urinary oxalate and glycollate. In one of them (GC), however, the plasma oxalate/creatinine ratio was only 0.031, and in the other (LC) before renal transplantation the ratio was only 0.042, these values being lower than any in Section 1 of the Table. This seems to indicate that the level of plasma oxalate is almost normal relative to that of creatinine despite marked hyperoxaluria and that the renal clearance of oxalate was some three times the clearance of creatinine. This must mean that oxalate was being generated in the kidneys and that most of it was secreted into the urinary tract with only a small fragment leaking back into the plasma. However, after renal transplantation with function of the newly grafted kidney the level of plasma creatinine fell to normal and the plasma oxalate/creatinine ratio rose to a mean of 0.165. This presumably means that the old kidneys ceased to excrete much urine when function was assumed by the new kidney, but nevertheless they continued to generate oxalate most of which then leaked back into the plasma to be handled in the same way by the transplanted kidney as it would be by a normal kidney. Thus it seems necessary to postulate that these two siblings generated the excess oxalate not in the liver as is usually the case in this condition, but in the kidney. Clearly more studies are needed to confirm this postulate, but if true it is of great importance for future management. Thus, while liver transplantation is now being advocated for primary hyperoxaluria (see Chap. 10), such a procedure would not be desirable for these two siblings in whom it would be better to remove their own kidneys and give them normal kidneys by transplantation. The idea that in primary hyperoxaluria there is an enzyme defect in the kidneys is certainly not a new one. Dean et al. (1966, 1967) found that the kidneys from two patients with primary hyperoxaluria

Table 1.1.   Plasma oxalate/creatinine ratios

Patient	Oxalate/ creatinine	SD	No of samples	Plasma creatinine ($\mu mol/l$)	Comments
1. Patients with primary hyperoxaluria					
AP	0.188	—	1	80	No response to high-dose pyridoxine
BG	0.182	0.025	4	71–82	10 years old
MW	0.180	—	2	46–60	2 years old
LS	0.127	0.007	56	64–551	Renal transplant did not affect ratio
IK	0.124	0.009	4	130–140	Post renal transplant. No response to pyridoxine
SK	0.120	—	1	45	6 years old. No response to pyridoxine
GF	0.113	0.026	86	74–1425	No response to high-dose pyridoxine. Renal transplant did not affect ratio
DW	0.110	0.042	35	120–1400	No response to high-dose pyridoxine. Renal transplant did not affect ratio
HR	0.107	—	3	261–305	Poor or no response to high-dose pyridoxine
RD	0.092	—	1	120	No response to pyridoxine
AR	0.089	—	1	105	On high dose pyridoxine
CC	0.089	—	1	135	
SS	0.069	—	3	110–115	Some but insufficient response to pyridoxine
AS	0.068	0.010	14	140–2100	Renal transplant did not affect ratio
RC	0.046	—	2	246–295	Good response to pyridoxine

2. *"Normal" renal failure*					
*	0.025	—	27	200–1700	
3. *"Renal hyperoxaluria"*					
GC	0.031	0.007	10	275–915	9 years old
LC	0.042	—	2	385–440	Pre-treatment. 7 years old
LC	0.1647	0.019	6	40–85	Post-treatment. 7 years old
4. *Normal subjects*					
PS	0.019	0.006	15		
AK	0.028	0.006	5		
PG	0.025	0.004	15		
CS	0.022	0.008	15		
PH	0.019	0.009	14		
GK	0.021	0.008	15		
GR	0.020	0.004	15		
MS	0.024	0.0003	3		
GH	0.033	—	1		7 years old
RH	0.043	—	1		8 months old

Mean normal ratio = 0.0224 ± 0.0032

Data from 99 samples on 10 normal subjects, 27 samples from chronic renal failure patients, 214 samples from 15 patients with primary hyperoxaluria 12 of whom had chronic renal failure.

failed to generate glycine from glyoxylate, while liver from the same patients did not show this defect. It may be therefore that the enzyme defect can occur either in liver or kidney in different patients with primary hyperoxaluria.

References

Allison MJ, Cook, HM, Milne DB (1986) Oxalate degradation by gastrointestinal bacteria from humans. J Hum Nutr 116: 455–460
Binder HJ (1974) Intestinal oxalate absorption. Gastroenterology 67: 441–446
Danpure CJ, Jennings PR, Watts RWE (1987) Enzymological diagnosis of primary hyperoxaluria type 1 by measurement of hepatic alanine: glyoxylate aminotransferase activity. Lancet I: 289–291
Dean BM, Griffin WJ, Watts RWE (1966) Primary hyperoxaluria. The demonstration of a metabolic abnormality in kidney tissue. Lancet I: 406
Dean BM, Watts RWE, Westwick WJ (1967) Metabolism of (1-^{14}C)glyoxylate, (1-^{14}C)glycollate, (1-^{14}C)glycine and (2-^{14}C)glycine by homogenates of kidney and liver tissue from hyperoxaluric and control subjects. Biochem J 105: 701–707
Finch AM, Kasidas GP, Rose GA (1981) Urine composition in normal subjects after oral ingestion of oxalate-rich foods. Clin Sci 60: 411–418
Herbert RW, Hirst EL, Percival EGU, Reynolds RJW, Smith F (1933) The constitution of ascorbic acid. J Chem Soc 1270–1290
Kasidas GP, Rose GA (1986) Measurement of plasma oxalate in healthy subjects and in patients with chronic renal failure using immobilised oxalate oxidase. Clin Chim Acta 157: 49–58
Kasidas GP, Rose GA (1987) The measurement of plasma oxalate and when this is useful. In: Vahlensieck W, Gasser G (eds) Pathogenese und Klinik der Harnsteine XI. Steinkopff-Verlag, Darmstadt, pp 143–147
Marshall RW, Cochran M, Hodgkinson A (1972) Relationships between calcium and oxalic acid intake in the diet and their excretion in the urine of normal and renal-stone-forming subjects. Clin Sci 43: 91–99
O'Regan P, Constable AR, Joekes AM, Kasidas GP, Rose GA (1980) Successful renal transplantation in primary hyperoxaluria. Postgrad Med J 56: 288–293
Rose GA (1985) Advances in analysis of urinary oxalate: the ascorbate problem solved. In: Schwille PO, Smith LH, Robertson WG, Vahlensieck W (eds) Urolithiasis and related clinical research. Plenum, New York, pp 637–644
Scheinman JI, Najarian JS, Mayer SM (1984) Successful strategies for renal transplantation in primary oxalosis. Kidney Int 25: 804–811
Watts RWE, Calne RY, Rolles K, Danpure CJ, Morgan SH, Mansell MA, Williams R, Purkiss P (1987) Successful treatment of primary hyperoxaluria type 1 by combined hepatic and renal transplant. Lancet II: 474–475
Wilson AE (1975) A report from ACS/WIH renal transplant registry. Renal transplantation in congenital and metabolic disease. JAMA 232: 148–153

Chapter 2

Assay of Oxalate and Glycollate in Urine

G. P. Kasidas

Introduction

Oxalate and glycollate are widely distributed in plant tissues (Hodgkinson 1977) where they both play important biochemical and physiological roles. In animals, oxalate presents toxicological problems associated with ingestion of large amounts of free acid (Zarembski and Hodgkinson 1967) and it assumes pathophysiological importance by virtue of the extreme insolubility, within the physiological range of pH, of its calcium salt in tissues and body fluids. The main impact of this is in relation to the urinary tract where crystals of calcium oxalate may be deposited to form calculi. The importance of oxalate in the formation of urinary calculi cannot be overstressed. Yet the rapid and precise measurement of this anion has presented considerable problems to investigators in this field. Some of these problems will be considered more fully in later sections of this chapter.

Glycollate is an intermediate in the photorespiratory carbon oxidation cycle (Chollet 1977) and is of great importance in the photosynthetic mechanism of plants. Unlike oxalate, it is not directly associated with the formation of urinary calculi but it relates with calcium oxalate urolithiasis because oxalate and glycollate arise from a common source, namely glyoxylate.

Relevant Chemistry of Oxalate and Glycollate Anions

Diagrammatic representations of the molecular structures of the oxalate and glycollate anions are shown in Fig. 2.1. Oxalate is the anion of a strong dicarboxylic acid which has a first dissociation constant of $pk_a^1 = 1.23$ and a

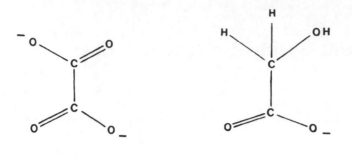

OXALATE   GLYCOLLATE

Fig. 2.1 Diagrammatic representations of the molecular structures of the oxalate and glycollate anions.

much weaker second dissociation constant of $pK_a^2 = 4.19$. Oxalate undergoes several reactions but of interest here are its oxidation to carbon dioxide plus water and its reduction to glycollate via glyoxylate. Figure 2.2 shows some of the reactions of oxalate which are relevant to this discussion.

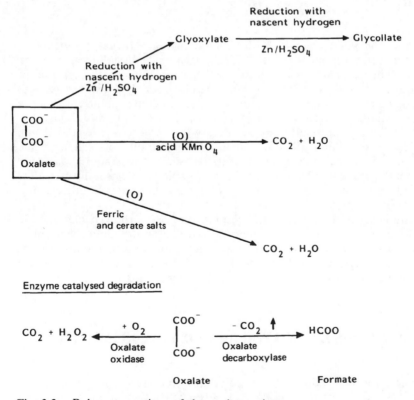

Fig. 2.2. Relevant reactions of the oxalate anion.

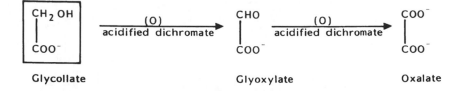

Glycollate   Glyoxylate   Oxalate

Glycollide formation

Like anions of other α hydroxyacids, two molecules of glycollate combine to form a six-membered condensation ring compound known as glycollide and this is readily reconverted to the original two molecules of glycollate by alkaline hydrolysis.

Enzymatic degradation

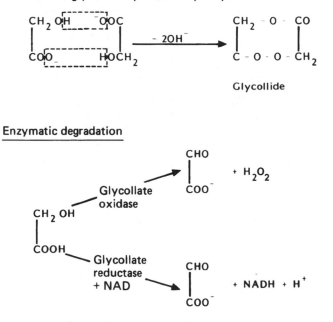

Fig. 2.3. Relevant reactions of the glycollate anion.

Glycollate is the anion of a monobasic α-hydroxyacid with a dissociation constant of $pK_a = 3.83$. Again the reactions of interest regarding this anion are shown in Fig. 2.3.

Assaying of Oxalate in Urine

Until recently the accurate measurement of oxalate in urine was a major problem. Numerous methods had been reported over the years and many discarded and replaced by others. Most of these earlier methods were prone

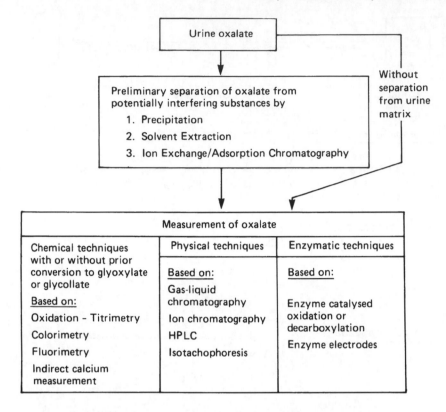

Fig. 2.4. Schematic diagram of the approaches used to measure urinary oxalate.

to errors, were time-consuming and required tedious sample-pretreatment procedures. Figure 2.4 outlines some of the main approaches used to assay oxalate in urine: the principles of these will be discussed below.

Methods for the Preliminary Separation of Oxalate from Other Urinary Constituents

The majority of methods for measuring oxalate in urine require preliminary separation of oxalate from potentially interfering substances. This is generally achieved either by precipitation of oxalate as its calcium, thorium, cerium, europium or lead salt (see below); by solvent extraction; or by ion exchange/adsorption chromatography. These methods for the separation of oxalate prior to analysis will be discussed below.

Separation by Precipitation

This has been the most commonly used of the separation methods for oxalate. Precipitation of oxalate as its calcium salt (Archer et al. 1957; Dick 1967) has been widely used. The main problems involved in this approach include incomplete precipitation due to the presence in the urine of inhibitors such as citrate and magnesium; losses of oxalate arising from washing and the coprecipitation of other components which disturb the subsequent estimation of the oxalate. To overcome some of the problems, various modifications have been introduced in attempts to minimise or correct for losses; these include the addition of a known amount of sodium oxalate (Koch and Strong 1969; Fraser and Campbell 1972) or the precipitation of calcium sulphate in the presence of ethanol at pH 7.0 to coprecipitate the calcium oxalate (Zarembski and Hodgkinson 1965; Hodgkinson and Williams 1972). The relatively lower solubilities of oxalate compounds of thorium-IV, cerium-III, europium-III and lead-II (Hodgkinson 1977) have led to their use with varying degrees of success in attempts to improve the completeness of precipitation. In some assays (Baadenhuisen and Jansen 1975; Koch and Strong 1969) radioisotopes have also been incorporated throughout the procedures to enable corrections to be made for any incomplete extraction or losses of oxalate. Improved accuracy was thus achieved at the expense of a more cumbersome method.

Separation by Solvent Extraction

Separation of oxalate, based on its partitioning between the aqueous phase of urine and an immiscible organic solvent such as diethyl ether (Yarbro and Simpson 1956; Hodgkinson and Zarembski 1961) or tri-*n*-butyl phosphate (Zarembski and Hodgkinson 1965) has been used in some assays for oxalate. Solvent extraction with tri-*n*-butyl phosphate is preferred because of the higher partition coefficient of this solvent system with respect to water.

Solvent extraction requires the subsequent precipitation of the oxalate and removal of the solvent prior to the determination of the separated oxalate. As a separation procedure it is not specific for oxalate; moreover, it is cumbersome and the conditions of extraction are critical.

Separation by Ion Exchange/Adsorption Chromatography

Oxalate has been separated by anion exchange chromatography (Chalmers and Watts 1972; Johansson and Tabova 1974) prior to its measurement. Separation by adsorption on to alumina is currently employed in the widely used Sigma assay (Sigma Diagnostics 1985). Here the oxalate in an acid urine (pH 2.5–3.0) is allowed to adsorb on to basic alumina in a batch-wise operation. After washing off the excess urine with acidified water ($\sim$ pH 3.0), the oxalate is eluted with 0.2M sodium hydroxide. Silica gel has also been used (Williams et al. 1979) to separate, by adsorption, the oxalate in urine, but with elution of the oxalate from the silica gel with a freshly prepared mixture of tert-amyl alcohol/chloroform/diethyl ether.

In both of these approaches (ion exchange and adsorption chromatography) the anions of other organic acids, eg. glyoxylate, glycollate, citrate, ascorbate, etc., invariably are separated with the oxalate and these have produced problems, as will be seen later, in the assaying of oxalate.

To overcome the limitations imposed in assays using preliminary separation procedures there has been a tendency towards the development of methods which directly measure the oxalate in unprocessed urine (see below).

Analytical Approaches for Determining Urinary Oxalate

The approaches for determining urinary oxalate can be categorised into three main groups although they overlap to some extent (see Fig. 2.4).

Chemical Techniques

Oxidation-Titrimetric Procedures. The oxalate from urine, after separation, has been oxidised by titration with hot acidified potassium permanganate (Archer et al. 1957; Dick 1975) according to the following equation:

$$5H_2C_2O_4 + 2MnO_4^- + 6H^+ \rightarrow 10CO_2 + 2Mn^{2+} + 8H_2O$$

The persistence of the pink colour of the potassium permanganate is used to determine the end-point of the titration. Difficulties can arise in the detection of the end-point, and in the standardisation of conditions such as temperature and titration rate. Also the presence of other reducing substances which inadvertently get separated with the oxalate during preliminary extraction procedures can yield erroneous titration values.

Cerium^{4+} ions have been used as an alternative oxidising agent but require an indicator (nitroferoin) to detect the end-point (Koch and Strong 1969; Blanka 1969). The reaction for this oxidation is shown thus:

$$H_2C_2O_4 + 2Ce^{4+} \rightarrow 2CO_2 + 2Ce^{3+} + 2H^+$$

It has been reported (Robertson and Rutherford 1980) that methods which used oxidation-titrimetry grossly underestimate the urinary oxalate.

Colorimetric and Fluorimetric Methods. Measurements of oxalate by colorimetry or fluorimetry have been the most popular methods used. Some of these require chemical reduction of the oxalate to glyoxylate or glycollate (see Fig. 2.2) prior to colorimetric or fluorimetric determinations.

A colorimetric procedure which does not require conversion of the oxalate to glyoxylate or glycollate has been reported by Baadenhuisen and Jansen (1975). Oxalate, previously extracted from urine, is used to inhibit the formation of a red complex between uranium-IV salt and 4-(2-pyridylazo)-resorcinol. Similarly fluorimetric assays not requiring the conversion of oxalate to either glyoxylate or glycollate were reported by Britton and Guyton (1969) and Gaetani et al. (1986). Both of these assays were based on the quenching by oxalate of the fluorescence of 1:1 zirconium–flavinol chelate in dilute sulphuric acid.

Colorimetric assays which require the reduction of oxalate to glyoxylate (Zarembski and Hodgkinson 1965) or glycollate (Dempsey et al. 1960; Hodgkinson and Zarembski 1961; Hodgkinson and Williams 1972) have been reported. The conditions for the reduction of oxalate to glyoxylate are very critical and these have been difficult to achieve (Zarembski and Hodgkinson

1965). The glyoxylate formed from the reduction of oxalate is allowed to react with phenylhydrazine/ferricyanide or hydrogen peroxide to form a red formazan which is then determined colorimetrically (Hamelle and Bressole 1975). Zarembski and Hodgkinson (1965) reported a fluorimetric assay in which the glyoxylate is reacted with resorcinol to form a highly fluorescent complex.

Other colorimetric methods had been developed which utilise reactions with glycollate and these have been preferred because of the difficulty in controlling the reduction of oxalate to glyoxylate. Oxalate is quantitatively reduced via glyoxylate to glycollate and to formaldehyde before reacting it with 2,7-dihydroxynaphthalene or 1,8-dihydroxynaphthalene-3-6-disulphonic acid (chromotropic acid) (Hodgkinson and Zarembski 1961; Hodgkinson and Williams 1972; Dempsey et al. 1960). As an alternative the oxalate had been converted to formate before reaction with indole to form a red complex which is determined colorimetrically (Hausman et al. 1956).

Limitations such as non-specificity and lack of sensitivity of chemical assays coupled with the disadvantages of preliminary separation of oxalate which have invariably been used in these procedures make these assays unsuitable for the accurate assay of oxalate in urine.

Indirect Technique Using Calcium Measurements. Calcium in urine can be measured accurately using atomic absorption spectrophotometry (Trudeau and Freier 1967; Gowans and Fraser 1986). Koehl and Abecassis (1976) utilised this fact and devised an assay for oxalate based on calcium determinations. They precipitated the oxalate in urine with a calcium salt and then measured the calcium, by atomic absorption spectrophotometry, in the supernatant or in the precipitate. The level of oxalate in the sample is calculated from the measured calcium. Again, errors can be introduced by the incomplete precipitation of calcium oxalate and the requirement, particularly at low concentrations of oxalate, to measure accurately small changes in relatively high concentrations of calcium.

Physical Techniques

Gas Chromatographic Procedures. Methods for measuring oxalate in urine based on gas chromatography have been reported (Chalmers and Watts 1972; Mee and Stanley 1973; Charransol et al. 1978; Dosch 1979; Yanagawa et al. 1983). Gas chromatography coupled with mass spectrometry (Mamer et al. 1971) or electron-capture techniques (Tocco et al. 1979) have also been used. The poor resolutions and inadequate sensitivity obtained by conventional packed columns led to the development of capillary gas-chromatographic procedures (Wolthers and Hayer 1982; Lopez et al. 1985).

The main disadvantage of these procedures is that they generally require preliminary separation of the oxalate followed by derivatisation to yield volatile esters for gas chromatographic separation and quantitation. High capital cost of equipment and tedium limit considerably the use of gas chromatography for measuring oxalate.

Ion Chromatographic Procedures. Measurement of oxalate in urine by ion chromatography has been reported (Robertson et al. 1982; Menon and Mahle

1983; Toyoda 1985). In this technique anions from diluted urine are allowed to bind to an anion-exchange resin and are eluted with alkaline potassium carbonate/bicarbonate buffer (pH 11.5–12.0). The eluate is then passed through a cation-exchange resin which removes the potassium ions, thus leaving the anions as free acids in a background of carbonic acid. The low conductivity of carbonic acid facilitates the quantitation of urinary anions including oxalate. The advantage is that the direct measurement of oxalate in the urine matrix is possible with this technique. However, during the alkaline stage ascorbate is converted to oxalate (see Fig. 2.8) although this can be prevented by the use of borate in the eluent (see below).

High Performance Liquid Chromatographic (HPLC) Procedures. Assays using HPLC separation of oxalate on a strong cation-exchange resin column with electrochemical detection (Mayer et al. 1979) or reverse phase separation with detection under ultraviolet light (Hughes et al. 1982) have been developed. Like the ion-chromatographic procedures, HPLC techniques can measure the oxalate in unprocessed urine (Larsson et al. 1982).

Isotachophoretic Procedures. Isotachophoretic separation of oxalate from other ionic species based on its net mobility in an electrical field within polytetrafluoroethylene (PTFE) capillaries followed by detection under ultraviolet light have been used in some assay procedures (Schmidt et al. 1980; Tschöpe et al. 1981; Schwendtner et al. 1982) for urinary oxalate. Some of these isotachophoretic procedures require preliminary separation of oxalate (Schmidt et al. 1980) whilst others do not require this step (Tschöpe et al. 1981).

Enzymatic Techniques

Because of the specificity offered by enzymes and the practical use of them without the need for preliminary separation, assays based on the enzyme-catalysed degradation of oxalate are now replacing the non-specific assays. Two enzymes, from a variety of sources (see below), have been used for the determination of urinary oxalate. These are oxalate decarboxylase (EC 4.1.1.3) and oxalate oxidase (EC 1.2.3.4) which respectively catalyse the decarboxylation and oxidation of oxalate (see Fig. 2.2).

The products derived from the enzyme-catalysed decomposition of oxalate have been monitored directly or after coupling with other standard detecting systems in assays for urinary oxalate. Thus the carbon dioxide formed by either of the two enzymes had been measured by manometry (Ribeiro and Elliot 1964), colorimetry (Knowles and Hodgkinson 1972) or changes in pH after trapping in a weak alkaline buffer (Hallson and Rose 1974; Kohlbecker et al. 1979) or conductivity measurements (Sallis et al. 1977; Bishop et al. 1982) or a radio-enzymatic technique (Bennett et al. 1978). Formate produced in the enzyme-catalysed decarboxylation has been measured after coupling with a second enzyme system such as the formate dehydrogenase/NAD system (Costello et al. 1976) or the formyl-tetrahydro-folate synthetase system (Jakoby 1974). With the former, problems arose from instability and contamination of the in-house preparation of formate dehydrogenase used in the assay (Costello et al. 1976). These were overcome by the use of commercial preparations of formate dehydrogenase (Boehringer Mannheim 1986; Urdal 1984). The different pH optima of oxalate decarboxylase ($\sim$ pH 3.5) and formate dehydrogenase ($\sim$ pH 7.0) and the requirement for further purification of the commercially available NAD+ to render it formate-free, presented some limitations in this assay (Costello et al. 1976).

The enzyme-catalysed oxidation of oxalate with oxalate oxidase yields hydrogen peroxide and carbon dioxide. The hydrogen peroxide can be determined photometrically with catalase/aldehyde dehydrogenase/NADP (Kohlbecker and Butz 1981) or in Trinder-type reactions (Barham and Trinder 1972) using peroxidase to couple oxidatively phenolic compounds to quinoneimine dyes (Potezny et al. 1983; Kasidas and Rose 1985a; Sigma Diagnostics 1985). In some assays (Kohlbecker et al. 1979) the carbon dioxide produced from the enzymatic oxidation of oxalate was measured by techniques described above.

Both of the oxalate-degrading enzymes have been used separately in assays developed at St Peter's Hospitals, London, to determine urinary oxalate. Hallson and Rose (1974) described a method for the direct measurement of oxalate in urine with the enzyme oxalate decarboxylase. Enzyme is added to urine at pH 3.5 in a sealed flask containing, in a separate compartment, weak alkaline buffer (see Fig. 2.5). Carbon dioxide generated from decarboxylation of the oxalate in the urine is trapped in the weak alkaline buffer to change its pH. The change in pH is related to the concentration of oxalate in the sample.

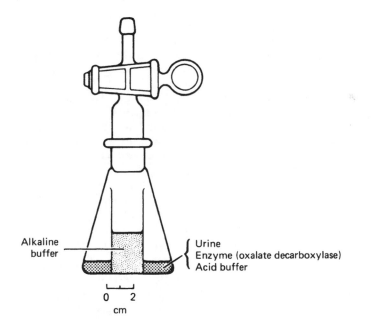

Alkaline buffer

Urine
Enzyme (oxalate decarboxylase)
Acid buffer

0   2
cm

Fig. 2.5. Apparatus for the enzymatic determination of oxalate with carbon dioxide measurement. Reproduced with permission from Hallson PC, Rose GA (1974) Clin Chim Acta 55: 29–39.

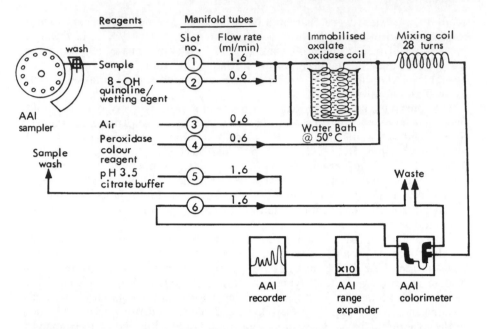

Fig. 2.6. Schematic diagram of the enzymatic continuous flow assay for urinary oxalate. Reproduced with permission from Kasidas GP, Rose GA (1985a); Ann Clin Biochem 22: 412–419.

This assay was used for a number of years to measure precisely urinary oxalate within the normal range. However, it suffered from drawbacks, requiring an overnight incubation, being labour-intensive and not being susceptible to automation.

For these reasons another assay using the second enzyme, oxalate oxidase, was developed in these laboratories (Kasidas and Rose 1985a). This second assay is automated and uses the enzyme immobilised on a nylon coil. A diagram illustrating the principles of this assay is shown in Fig. 2.6. This new assay has several advantages over the other assays described so far. These include:

1. Preliminary separation of oxalate from urine is not required. The assay is specific and sensitive (detection limit 0.5 μmol/l when slightly adapted for the measurement of plasma oxalate (see Chap. 4)
2. Improved performance characteristics from automation and reusability of the enzyme reactor
3. Cost effectiveness from automation, reusability of the enzyme (cost of reagent per assay £0.10) (Kasidas and Rose 1985a)
4. Avoidance of ascorbate interference (see below) with buffered sodium nitrite and acidic (pH 3.5) conditions of assay
5. Use of the same equipment for plasma oxalate determinations. (See Chap. 4)

Oxalate oxidase from barley seedlings is used in the above assay but enzyme extracted from other sources including moss (Laker et al. 1980), beet stems (Obzansky and Richardson 1983), sorghum seeds (Pundir et al. 1985) and unripe banana peel (Inamdar et al. 1986; Raghavan and Devasagayam 1985) have also been used. Inamdar et al. (1986) reported a manual procedure for urinary oxalate in which the reusability of oxalate oxidase was achieved in a novel way. They stabilised banana-peel oxalate oxidase by raising a rat–banana-oxalate-oxidase immune complex which was then used for analysis of oxalate.

Commercial Enzymatic Kits. Commercial enzymatic kits have been introduced (Sigma Diagnostics 1985; Boehringer Mannheim Biochemica 1982) and these are gaining acceptance in many clinical chemistry laboratories where small numbers of urinary oxalate assays are requested. Although expensive, they become less so when adapted for use on discrete chemical analysers (Bradley et al. 1987; Sigma Diagnostics 1985; Urdal 1984) but are not entirely free from problems (Barlow 1987; Parkinson et al. 1987; Kasidas and Rose 1987; Glick 1987 and see above and below). Additionally, because of the requirement to adsorb the oxalate on to alumina in the Sigma kit, errors can be introduced (see p. 11) and impose restrictions on the accuracy of the results (Parkinson et al. 1987; Barlow 1987).

Enzyme Electrodes. Oxalate-specific enzyme electrodes have been developed to measure levels of oxalate in urine (Kobos and Ramsey 1980; Nabi Rahni et al. 1986; Vadgama et al. 1984). Kobos and Ramsey (1980) and Vadgama et al. (1984) used immobilised or entrapped oxalate decarboxylase combined to a carbon-dioxide-sensor electrode while Nabi Rahni et al. (1986) used oxalate oxidase combined to an oxygen electrode to determine levels of oxalate. These electrodes still require further development to render them suitable for routine laboratory analysis. The main problem has been the slow response-time and the lack of sensitivity.

Ascorbate – a Major Problem in the Assaying of Oxalate in Urine

Apart from the problems encountered in oxalate methodologies (discussed above), another special problem which had been previously overlooked concerns ascorbate, a common constituent of urine. In 1933, it was reported that ascorbate could oxidise spontaneously to oxalate (Herbert et al. 1933) probably by the reaction route shown in Fig. 2.7. This important fact, overlooked for fifty years, was restated simultaneously by Kasidas and Rose (1985b) and Mazzachi et al. (1985). It has since been documented by other investigators (Glick 1987; Chalmers et al. 1985; Crawford et al. 1985; Parkinson et al. 1987) that the conversion of oxalate can occur during analysis and post-sample-collection periods. The conversion of ascorbate to oxalate occurs more quickly at alkaline pH values with further enhancement by charcoal (Fig. 2.8) (Kasidas and Rose 1985b). Assays which use alkaline conditions can generate oxalate from ascorbate. The generation of oxalate from ascorbate in the ion-chromatographic and Sigma oxalate assays is shown in Fig. 2.9.

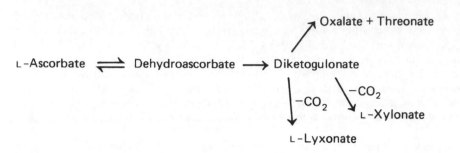

Fig. 2.7. Reaction route for the conversion of ascorbate to oxalate. Reproduced with permission from Kasidas GP, Rose GA (1985a); Ann Clin Biochem 22: 412–419.

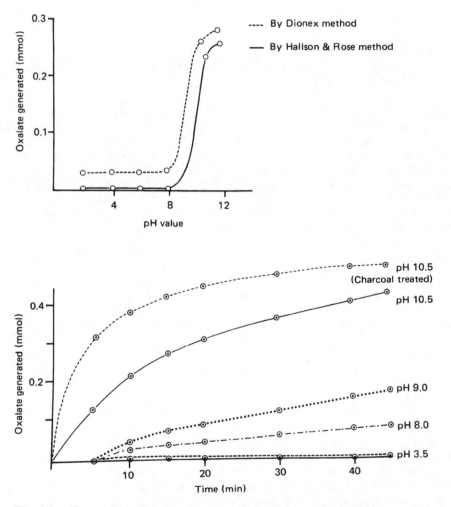

Fig. 2.8. Generation of oxalate from ascorbate (0.5 mmol/l) at different pH values.

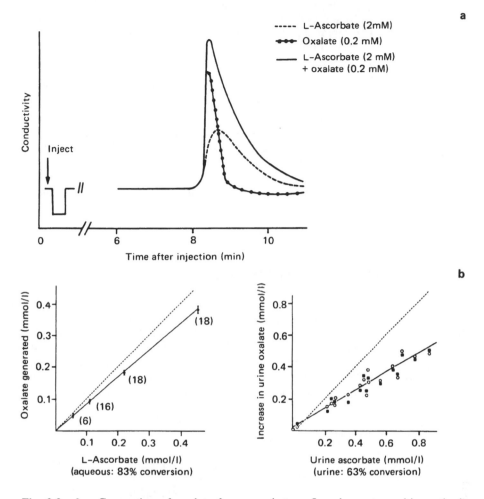

Fig. 2.9a, b.   Generation of oxalate from ascorbate. **a** Ion chromatographic method. **b** Sigma urine oxalate procedure.

Measures have been taken to suppress the interference by ascorbate in a number of methods. Thus, in the ion-chromatographic procedure boric acid can be incorporated in the eluent (Scurr et al. 1985). In an enzymatic-rate-analysis assay (Chalmers et al. 1985) ethylenediaminetetracetic acid (EDTA) is used; sodium nitrite, buffered at pH 3.5, is effective in the automated immobilised oxalate-oxidase assay (Kasidas and Rose 1985a) and in the Sigma assay (Kasidas and Rose 1987). Acidic ferric chloride has been advocated (Obzansky and Richardson 1983; Kasidas and Rose 1985b; Rose 1985; Crider 1985) but its effect has not been consistent in the Sigma assay (Barlow 1987; Kasidas and Rose 1987) and it can lead to a reduced life-span of the columns in the ion-chromatographic procedure (Scurr et al. 1985). Preincubation with ascorbate oxidase (EC 1.10.3.3) in solution (Crawford et al. 1985; Ichiyama et al. 1985) or in an immobilised state in the form of an ascorbate-oxidase spatula (Boehringer Mannheim Biochemica 1986) has been used.

It is therefore incumbent to show in any new assay for oxalate that there is no interference from ascorbate.

Normal Range for Urinary Oxalate

With the advent of specific and reliable methods for measuring urinary oxalate and with precautions taken to eliminate the problems discussed above, there is now better agreement on the upper level of the normal range. At St Peter's Hospitals the upper level of the normal range for urinary excretion of oxalate is considered to be 0.45 mmol/24 h. Hyperoxaluria ensues when the level of urinary oxalate rises above this value.

Assaying for Urinary Glycollate

Unlike oxalate, the measurement of glycollate in urine has received relatively little attention. The presence of glycollate in urine was first demonstrated by Nordmann et al. (1954). Some of the earlier techniques for urinary glycollate used paper and silica-gel chromatography and are only semi-quantitative (Nordmann et al. 1954; Osteux and Laturaze 1956; Meites 1957). Quantitative methods for determining glycollate in urine are presented below.

Chemical Techniques

Chemical methods based on the formation of a coloured complex between formaldehyde, produced by the reduction of glycollate, and 1,8-dihydroxynaph-thalene-3,6-disulphonic acid (chromotropic acid) (Hodgkinson and Zarembski 1961; Chernoff and Richardson 1978) have been described. In some assays the chromotropic acid has been substituted with 2,7-dihydroxynaphthalene (Chow et al. 1978) or β naphthol (Viccaro and Ambyne 1972). Glycollate is reduced in the above procedures by heating it with concentrated sulphuric acid. As with the techniques for determining oxalate, preliminary separation of glycollate from other urinary constituents is necessary before assaying, and this step can introduce errors. Furthermore, any contaminant separated with glycollate which can produce formaldehyde on heating with concentrated sulphuric acid will also introduce errors.

Isotope Dilution Procedures

Isotope dilution methods for urinary glycollate have been described (Hockaday et al. 1965; Johannson and Tabova 1974) but only in combination with colorimetric procedures which themselves are unreliable.

Physical Techniques

Gas–liquid chromatographic methods have been developed for the simultaneous estimations of organic acids (Chalmers and Watts 1972; Ferraz and Relvas 1965). Preliminary isolation of glycollate followed by subsequent derivatisation to its methyl or other esters have been necessary before passing it through a chromatographic column for analysis. Gas chromatographic techniques for urinary glycollate are not entirely problem-free and, moreover, they require the use of relatively expensive equipment which is not easily accessible to the majority of laboratories. A method (Hewlett et al. 1983) using HPLC and detection with ultraviolet light has been reported but this also required extraction of the glycollate with methyl-ethyl-ketone and derivatisation with *o*-*p*-nitrobenzyl-N, N^1-diisopropylisourea prior to analysis.

Enzymatic Techniques

The oxidation of glycollate to glyoxylate and hydrogen peroxide can be enzymatically catalysed by glycollate oxidase ((S)-2-hydroxy-acid oxidase) (EC 1.1.3.15)) (See Fig. 2.3). An assay using glycollate oxidase extracted from rat liver has been reported (Metzger et al. 1975), but it suffers from drawbacks such as the use of an inconvenient source of the glycollate oxidase (rat liver) and the use of a non-specific chemical measurement of the formed glyoxylate. At St Peter's Hospitals an enzymatic assay using spinach glycollate oxidase had been developed (Kasidas and Rose 1979) for the measurement of glycollate

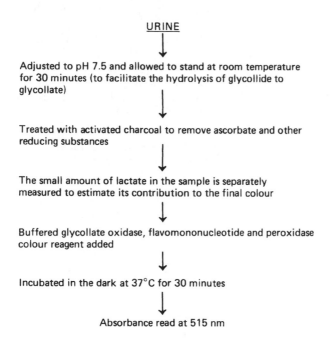

URINE

Adjusted to pH 7.5 and allowed to stand at room temperature for 30 minutes (to facilitate the hydrolysis of glycollide to glycollate)

Treated with activated charcoal to remove ascorbate and other reducing substances

The small amount of lactate in the sample is separately measured to estimate its contribution to the final colour

Buffered glycollate oxidase, flavomononucleotide and peroxidase colour reagent added

Incubated in the dark at 37°C for 30 minutes

Absorbance read at 515 nm

Fig. 2.10. Principles of the manual enzymatic assay for urinary glycollate.

in urine. In this assay certain interfering substances, but not lactate, are removed by adsorption on to charcoal. Spinach glycollate oxidase catalyses the oxidation of glycollate to produce hydrogen peroxide which is subsequently quantitated in a second colorimetric enzyme system with peroxidase and 4-aminophenazone. Even though at a slower rate, lactate is also oxidised by glycollate oxidase to produce hydrogen peroxide. The lactate can be measured separately and its contribution to the final absorbance can be calculated and then subtracted.

This technique has been applied to a discrete centrifugal analyser and continuous flow autoanalyser system (Bais et al. 1985). Glycollate oxidase, immobilised on to the inner surface of nylon tubing, was used in the continuous flow system but unfortunately the enzyme is not stable enough in the immobilised state for routine use (life span ≃ 12 days: Bais et al. 1985).

The procedure for the manual enzymatic assay for glycollate is summarised in Fig. 2.10. At St Peter's Hospitals 0.33 mmol/24 h is taken as the upper limit of the normal urinary range. However, there does seem to be some variation in quoted normal ranges and local assessment of the reference range seems advisable.

Usefulness of Urinary Glycollate in Determining the Origin of Hyperoxaluria

It has been normal practice at St Peter's Hospitals to measure the glycollate levels in all urine samples which show hyperoxaluria. Urinary glycollate has been shown here to be raised in all types of metabolic hyperoxaluria which are mediated through glyoxylic acid (see Table 2.1). These include type 1 primary hyperoxaluria (Chap. 5), mild metabolic hyperoxaluria (Chap. 8), hyperoxaluria arising from over-absorption of glycine in transurethral resection of the prostate (Fitzpatrick et al. 1982), ethylene-glycol intoxication (Hewlett et al. 1983) and pyridoxine deficiency (Gershoff 1964). When hyperoxaluria

Table 2.1. Differentiation of the origin of hyperoxaluria based on urinary glycollate level

Normal glycollate	Raised glycollate
Non-metabolic hyperoxaluria	All forms of metabolic hyperoxaluria which are mediated through glyoxylate
Hyperabsorption of exogenous oxalate as in steatorrhoea, idiopathic hypercalciuria and therapy with cellulose phosphate	Type 1 primary hyperoxaluria
	Mild metabolic hyperoxaluria
Excessive dietary intake of oxalate	Overabsorption of glycine in TUR prostatectomy
Ascorbate and purine induced hyperoxaluria	Ethylene glycol intoxication
[a]Methoxyflurane	Pyridoxine deficiency

[a]Glycollate excretion never reported but presumed absent

arises non-metabolically from excessive dietary intake of oxalate or hyperabsorption of exogenous oxalate as in steatorrhoea, hyperglycollaturia does not occur (Rampton et al. 1979). The hyperoxaluria arising from ascorbate and purine degradation also do not show hyperglycollaturia (Kasidas 1980).

Urinary glycollate can therefore be used (Rose and Kasidas 1979) to distinguish exogenously mediated non-metabolic hyperoxaluria from the metabolic forms of hyperoxaluria arising via the glyoxylate route.

Summary

1. Assays are now available for the reliable and routine measurement of oxalate in urine without requiring any additional preliminary separation procedures

2. Means are also available for the removal of interference by ascorbate in oxalate assays thus reducing the risk of finding factitious hyperoxaluria

3. The detection of hyperoxaluria, even mild forms, is possible and can now be made with a greater degree of confidence

4. Recognition of the cause of hyperoxaluria is assisted by the measurement of urinary glycollate

5. The effect of treatment of hyperoxaluria can be followed by assaying the levels of oxalate and glycollate anions in urine

References

Archer HE, Dormer AE, Scowen EF, Watts RWE (1957) Studies on the urinary excretion of oxalate by normal subjects. Clin Sci 16: 405–411
Baadenhuisen H, Jansen AP (1975) Colorimetric determination of urinary oxalate recovered as calcium oxalate. Clin Chim Acta 62: 315–324
Bais R, Nairn JM, Potezny N, Rofe AM, Conyers RAJ, Bär A (1985) Urinary glycollate measured by use of S-2-hydroxyacid oxidase. Clin Chem 31: 710–713
Barham D, Trinder P (1972) An improved colour reagent for the determination of blood glucose by the oxidase system. Analyst 97: 142–145
Barlow IM (1987) Obviating interferences in the assay of urinary oxalate. Clin Chem 33: 855–858
Bennett DJ, Cole FE, Frohlich ED, Erwin DT (1978) Radioenzymatic procedure for urinary oxalate determination. J Lab Clin Med 91: 822–830
Bishop M, Freudiger H, Largiader U, Sallis JD, Felix R, Fleisch H (1982) Conductivometric determination of urinary oxalate with oxalate decarboxylase. Urol Res 10: 191–194
Blanka B (1960) Determination of oxalic acid in urine. Vnitr Lek 6: 1301–1309
Boehringer Mannhein Biochemica (1986) Oxalic acid. UV-method. In: Methods of biochemical analysis and food analysis. Boehringer Mannhein GmBH Biochemica, pp 86–88
Bradley CA, Aleshire SL, Parl FF (1987) Automated enzymatic method for measuring oxalate in urine. Clin Chem 33: 1076–1077
Britton DA, Guyton JC (1969) Fluorimetric determination of oxalate ion. Anal Chim Acta 44: 397–401
Chalmers AH, Cowley DM, McWhinney BC (1985) Stability of ascorbate in urine; relevance to analysis for ascorbate and oxalate. Clin Chem 31: 1703–1705
Chalmers RA, Watts RWE (1972) The quantitative extraction and gas liquid chromatographic determination of organic acids in urine. Analyst 97: 958–967
Charransol G, Barthelemy CH, Desgrez P (1978) Rapid determination of urinary oxalic acid by gas-liquid chromatography without extraction. J Chromatogr 145: 452–455

Chernoff HN, Richardson KE (1978) The effect of endogenous L-phenyllactate on oxalate, glycollate, and glyoxylate excretion by phenylketonuric subjects. Clin Chim Acta 83: 1–6

Chollet R (1977) The biochemistry of photorespiration. Trends in Biochemical Sciences 2: 155–159

Chow FC, Hamar DW, Boulay JP, Lewis LD (1978) Prevention of oxalate urolithiasis by some compounds. Invest Urol 15: 493–495

Costello J, Hatch M, Bourke E (1976) An enzymatic method for the spectrophotometric determination of oxalic acid. J Lab Clin Med 87: 903–908

Crawford GA, Mahony JF, Györy AZ (1985) Measurement of urinary oxalate in the presence of ascorbic acid. Clin Chim Acta 147: 51–57

Crider QE (1985) Effect of diluting samples for enzymatic determination of urinary oxalate. Clin Chem 31: 1080–1081

Dempsey EF, Forbes AP, Melick RA, Henneman PH (1960) Urinary oxalate excretion. Metabolism 9: 52–58

Dick M (1967) A simplified urinary oxalate method. Proc Assoc Clin Biochem 4: 186–187

Dosch W (1979) Rapid and direct gas chromatographic determination of oxalic acid in urine. Urol Res 7: 227–234

Ferraz FGP, Relvas ME (1965) Séparation et identification par chromatographie sur colonne et ou phase gazeuse des acides organiques des milieux biologiques. Clin Chim Acta 11: 244–248

Fitzpatrick JM, Kasidas GP, Rose GA (1981) Hyperoxaluria following glycine irrigation for transurethral prostatectomy. Br J Urol 53: 250–252

Fraser J, Campbell DJ (1972) Indirect measure of oxalate acid in urine by atomic absorption spectrophotometry. Clin Biochem 5: 99–103

Gaetani E, Laureri CF, Vitto M, Borghi L, Elia GF, Novarini A (1986) Determination of oxalate in urine by flow injection analysis. Clin Chim Acta 156: 71–76

Gershoff SN (1964) Vitamin B_6 and oxalate metabolism. In: Aurbach GD, McCormic DB (eds) Vitamins and hormones. Academic Press, New York, 22: 581–589

Glick JH (1987) Deficiencies of Sigma Diagnostics urinary oxalate method in the presence of ascorbate. Clin Chem 33: 419–420

Gowans EMS, Fraser CG (1986) Five methods for determining urinary calcium compared. Clin Chem 32: 1560–1562

Hallson PC, Rose GA (1974) A simplified and rapid enzymatic method for determination of urinary oxalate. Clin Chim Acta 55: 29–39

Hamelle, G, Bressole F (1975) Microdosage de l'acide oxalique dans les urines. Travaux de la Société de Pharmacie de Montpellier 35: 195–204

Hausman ER, McAnally JS, Lewis GT (1956) Determination of oxalate in urine. Clin Chem 2: 439–444

Herbert RW, Hirst EL, Percival EGV, Reynolds RJW, Smith F (1933) The constitution of ascorbic acid. J Chem Soc 1270–1290

Hewlett TP, Ray AC, Reagor JC (1983) Diagnosis of ethylene glycol (anti-freeze) intoxication in dogs by determination of glycolic acid in serum and urine with high pressure liquid chromatography and gas chromatography-mass spectrometry. J Assoc Off Anal Chem 66: 276–283

Hockaday TDR, Frederick EW, Clayton JE, Smith LH (1965) Studies on primary hyperoxaluria, II Urinary oxalate, glycollate and glyoxylate measurement by isotope dilution methods. J Lab Clin Med 65: 677–687

Hodgkinson A (1977) Oxalic acid in biology and medicine. Academic Press, New York

Hodgkinson A, Williams A (1972) An improved colorimetric procedure for urine oxalate. Clin Chim Acta 36: 127–132

Hodgkinson A, Zarembski PM (1961) The determination of oxalic acid in urine. Analyst 86: 16–21

Hughes H, Hagen L, Sutton RAL (1982) Determination of urinary oxalate by high performance liquid chromatography. Anal Biochem 119: 1–3

Ichiyama A, Nakai E, Funai T, Oda T, Katafuchi R (1985) Spectrophotometric determination of oxalate in urine and plasma with oxalate oxidase. J Biochem (Tokyo) 98: 1375–1385

Inamdar KV, Tarachand U, Davasagayam TPA, Raghavan KG (1986) Immune complex of banana oxalate oxidase: use in quantitation of urinary oxalate. Anal Lett 19: 1987–1999

Jakoby WB (1974) Oxalate; formate. In: Bergmeyer HU (ed) Methods of enzymatic analysis (2nd Edition). Academic Press, New York 3: 1542–1548

Johansson S, Tabova R (1974) Determination of oxalic and glycollic acids with isotope dilution methods and studies on the determination of glyoxylic acid. Biochem Med 11: 1–9

Kasidas GP (1980) Factors affecting the levels of oxalate and glycollate in the urine and plasma of normal subjects and calcium oxalate stone formers. PhD Thesis, University of London

Kasidas, GP, Rose GA (1979) A new enzymatic method for the determination of glycollate in urine and plasma. Clin Chim Acta 96: 25–36

Kasidas GP, Rose GA (1985a) Continuous flow assay for urinary oxalate using immobilised oxalate oxidase. Ann Clin Biochem 22: 412–419

Kasidas GP, Rose GA (1985b) Spontaneous generation of oxalate from L-ascorbate in some assays for urinary oxalate and its prevention. In: Schwille PO, Smith LH, Robertson WG, Vahlensieck W (eds) Urolithiasis and related clinical research. Plenum Press, New York, pp. 653–656

Kasidas GP, Rose GA (1987) Removal of ascorbate from urine prior to assaying with a commercial oxalate kit. Clin Chim Acta 164: 215–221

Knowles CF, Hodgkinson A (1972) Automated enzymic determination of oxalic acid in human serum. Analyst 97: 474–481

Kobos RK, Ramsey TA (1980) Enzyme electrode system for oxalate determination utilising oxalate decarboxylase immobilised on a carbon dioxide sensor. Anal Chim Acta 121: 111–118

Koch GH, Strong FM (1969) Determination of oxalate in urine. Anal Biochem 27: 162–171

Koehl C, Abecassis J (1976) Determination of oxalic acid in urine by atomic absorption spectrophotometry. Clin Chim Acta 70: 71–77

Kohlbecker G, Butz M (1981) Direct spectrophotometric determination of serum and urinary oxalate with oxalate oxidase. J Clin Chem Clin Biochem 19: 1103–1106

Kohlbecker G, Richter L, Butz M (1979) Determination of oxalate in urine using oxalate oxidase: comparison with oxalate oxidase. J Clin Chem Clin Biochem 17: 309–313

Laker MF, Hofmann AF, Meeuse BJD (1980) Spectrophotometric determination of urinary oxalate with oxalate oxidase prepared from moss. Clin Chem 26: 827–830

Larsson L, Libert B, Asperud M (1982) Determination of urinary oxalate by reversed-phase ion pair high performance liquid chromatography. Clin Chem 28: 2272–2274

Lopez M, Tuchman M, Scheinman JI (1985) Capillary gas chromatography measurement of oxalate. Kidney Int 18: 82–84

Mamer OA, Crawhall JC, Tjoa SS (1971) The identification of urinary acids by coupled gas chromatography-mass spectrometry. Clin Chim Acta 32: 171–184

Mayer WJ, McCarthy JP, Greenberg MS (1979) The determination of oxalic acid in urine by high performance liquid chromatography with electrochemical detection. J Chromatogr Sci 17: 656–660

Mazzachi BC, Teubner JK, Ryall RL (1985) The effect of ascorbic acid on urine oxalate measurement. In: Schwille PO, Smith LH, Robertson WG, Vahlensieck W (eds) Urolithiasis and related clinical research. Plenum Press, New York, pp. 649–652

Mee JM, Stanley RW (1973) A rapid gas-liquid chromatographic method for determining oxalic acid in biological materials. J Chromatogr 75: 242–243

Meites S (1957) Partition chromatography of organic acids in body fluids with silica gel; application of the method to normal human urine. Clin Chem 3: 263–269

Menon M, Mahle CJ (1983) Ion chromatographic measurement of oxalate in unprocessed urine. Clin Chem 29: 369–371

Metzger RP, Sauerheber RD, Lyons SA, Westall JR (1975) The effect of Streptozotocin diabetes on the levels of glycollate and lactate excreted in rat urine. Arch Biochem Biophys 169: 555–559

Nabi Rahni MA, Guilbault GG, de Olivera NG (1986) Immobilised enzyme electrode for the determination of oxalate in urine. Anal Chem 58: 523–526

Nordmann R, Gauchery O, du Ruisseau JP, Thomas Y, Nordmann J (1954) Chromatographie sur papier des acides organiques non volatiles des liquides biologiques II Le chromatogram qualitatif de l'urine humaine normal. Bull Soc Chim Biol 36: 1641–1654

Obzansky DM, Richardson KE (1983) Quantification of urinary oxalate with oxalate oxidase from beet stems. Clin Chem 29: 1815–1819

Osteux R, Laturaze J (1956) Chromatographie sur papier des acides organiques non volatile de l'urine humaine. Clin Chim Acta 1: 379–396

Parkinson IS, Sheldon WL, Laker MF, Smith PA (1987) Critical evaluation of a commercial enzyme kit (Sigma) for determining oxalate concentrations in urine. Clin Chem 33: 1203–1207

Potezny N, Bais R, O'Loughlin PD, Edwards JB, Rofe AM, Conyers RAJ (1983) Urinary oxalate determination by use of immobilised oxalate oxidase in a continuous flow system. Clin Chem 29: 16–20

Pundir CS, Nath R, Thind SK (1985) Enzymatic assay of oxalate using oxalate oxidase from sorghum leaves. In: Schwille PO, Smith LH, Robertson WG, Vahlensieck W (eds) Urolithiasis and related clinical research. Plenum Press, New York, pp 657–660

Raghavan KG, Devasagayam TPA (1985) Oxalate oxidase from banana peel for determination of urinary oxalate. Clin Chem 31: 649

Rampton DS, Kasidas GP, Rose GA, Sarner M (1969) Oxalate loading test: a screening test for steatorrhoea. Gut 20: 1089–1094

Ribeiro ME, Elliot JS (1964) Direct enzymatic determination of urinary oxalate. Invest Urol 2: 78–81

Robertson WG, Rutherford A (1980) Aspects of the analysis of oxalate in urine—a review. Scand J Urol Nephrol [Suppl] 53: 85–93

Robertson WG, Scurr DS, Smith A, Orwell RL (1982) The determination of oxalate in urine and urinary calculi by a new ion-chromatographic technique. Clin Chim Acta 126: 91–99

Rose GA (1985) Advances in analysis of urinary oxalate; the ascorbate problem solved. In: Schwille PO, Smith LH, Robertson WG, Vahlensieck W (eds) Urolithiasis and related clinical research. Plenum Press, New York, pp 634–644

Rose GA, Kasidas GP (1979) New enzymatic method for measurement of plasma and urinary glycollate and its diagnostic value. In: Gasser G, Vahlensieck W (eds) Pathogenese und Klinik der Harnsteine VII. Steinkopff, Darmstadt, pp 252–260

Sallis JD, Lumley MF, Jordan JE (1977) An assay of oxalate based on the conductometric measurement of enzyme-liberated carbon dioxide. Biochem Med 18: 371–377

Schmidt K, Hagmaier V, Bruchelt G, Rutishauer G (1980) Analytical isotachophoresis: a rapid and sensitive method for determination of urinary oxalate. Urol Res 8: 177–180

Schwendtner N, Achilles W, Engelhardt W, Schwille PO, Sigel A (1982) Determination of urinary oxalate by isotachophoresis. Practical improvement and critical evaluation. J Clin Chem Clin Biochem 20: 833–836

Scurr DS, Januzovich N, Smith A, Sargeant VJ, Robertson WG (1985) A comparison of three methods for measuring urinary oxalate—with a note on ascorbic acid interference. In: Schwille PO, Smith LH, Robertson WG, Vahlensieck W (eds) Urolithiasis and related clinical research. Plenum Press, New York, pp 645–648

Sigma Diagnostics (1985) Oxalate—quantitative enzymatic determination in urine at 590 nm. In: Sigma Technical Bulletin 590. Sigma Chemical Co, Poole, Dorset, UK, pp 1–11

Tocco DJ, Duncan AEW, Noll RM, Duggan DE (1979) An electron-capture gas chromatographic procedure for the estimation of oxalic acid in urine. Anal Biochem 94: 470–476

Toyoda M (1985) A simple ion-chromatographic method for determination of urinary oxalate. Urol Res 13: 179–183

Trudeau DL, Freier EF (1967) Determination of calcium in urine and serum by atomic absorption spectrophotometry. Clin Chem 13: 101–114

Tschöpe W, Brenner R, Ritz E (1981) Isotachophoresis for the determination of oxalate in unprocessed urine. J Chromatogr 222: 41–52

Urdal P (1984) Enzymic assay for oxalate in unprocessed urine as adapted for a centrifugal analyser. Clin Chem 30: 911–913

Vadgama P, Sheldon W, Guy JM, Covington AK, Laker MF (1984) Simplified urinary oxalate determination using an enzyme electrode. Clin Chim Acta 142: 193–201

Viccaro JP, Ambyne EL (1972) Colorimetric determination of glycollic acid with β-Naphthol. Microchem J 17: 710–718

Williams VP, Ching DK, Cederbaum SD (1979) Adsorption of organic acids from amniotic fluid and urine on to silica gel before analysis by gas chromatography and combined gas chromatography/mass spectrometry. Clin Chem 25: 1814–1820

Wolthers BG, Hayer M (1982) The determination of oxalic acid in plasma and urine by means of capillary gas chromatography. Clin Chim Acta 120: 87–102

Yanagawa M, Ohkawa H, Tada S (1983) The determination of urinary oxalate by gas chromatography. J Urol 129: 1163–1165

Yarbro CL, Simpson RE (1956) The determination of total urinary oxalate. J Lab Clin Med 48: 304–310

Zarembski PM, Hodgkinson A (1967) Plasma oxalic acid and calcium levels in oxalate poisoning. J Clin Path 20: 283–285

Zarembski PM, Hodgkinson A (1965) The fluorimetric determination of oxalic acid in blood and other biological materials. Biochem J 96: 712–721

Chapter 3

Experiences with an External Quality Assessment Scheme for Urinary Oxalate

C. T. Samuell

Introduction

One of the recent advances in our understanding of calcium oxalate urolithiasis has been the recognition of the importance of hyperoxaluria in the pathogenesis of this large and growing problem (Robertson and Nordin 1969; Robertson et al. 1979; Robertson and Peacock 1980; Baggio et al. 1983; Antonacci et al. 1985; Jaeger et al. 1985). It must now be accepted that hyperoxaluria, albeit often mild, is a more important finding than hypercalciuria in many patients and indeed the presence of hypercalciuria should not prevent a thorough search for coexisting hyperoxaluria. Thus there is clearly a need for urinary oxalate to be estimated in all patients presenting with calcium oxalate stones and those laboratories serving clinics dealing with such patients should anticipate a growing demand for the precise and accurate assay of urinary oxalate.

Urinary Oxalate Assay – The Need for External Quality Assessment

For the reasons stated above many clinical chemistry departments may be faced with setting up an assay for urinary oxalate or reviewing their existing methodology. There have been various approaches to the measurement of urinary oxalate and the subject has been reviewed elsewhere (Hodgkinson 1977; Robertson and Rutherford 1980; Zerwekh et al. 1983). It is thought that many of the older precipitation procedures are probably inadequate for demonstrating the minor elevations of oxalate excretion seen in many calcium oxalate-stone formers, having traditionally been used to detect the very high urine-oxalate levels found in primary hyperoxaluria. Perhaps the most significant

Table 3.1.   Some of the factors suggesting the need for an external quality assessment scheme

i	Recognition of the importance of mild hyperoxaluria in the pathogenesis of calcium-oxalate urolithiasis
ii	Availability of commercial kits for oxalate assay
iii	Discrepancy between results obtained at St Peter's Hospitals and those reported from referring hospitals
iv	Local expertise with several techniques and their problems

advance in oxalate measurement has been the development of enzymatic procedures utilising the activity of either oxalate decarboxylase (Hallson and Rose 1974; Costello et al. 1976; Chalmers and Cowley 1984) or oxalate oxidase (Bais et al. 1980; Laker et al. 1980; Kohlbecker 1981; Buttery et al. 1983; Obzansky and Richardson 1983). Assays based on both of these enzymes are now available as commercial kits and their convenience obviously makes them attractive to the laboratory with a relatively small number of measurements of urinary oxalate to perform. Indeed, as will be shown below, the colorimetric kit based on oxalate oxidase and marketed by Sigma Chemical Co Ltd is currently the most widely used method in the UK. Fully automated procedures based on immobilised oxalate oxidase have also been described (Bais et al. 1980; Kasidas and Rose 1985a). In addition, more sophisticated techniques such as ion chromatography (Mahle and Menon 1982) and high pressure liquid chromatography (HPLC) (McWhinney et al. 1986) are being applied to the problem although, for obvious reasons, they may have a limited role in the routine laboratory. All of the above techniques may have their limitations but the potential problem of ascorbate interference in many methods has recently been highlighted (Buttery et al. 1983; Mazzachi et al. 1984; Kasidas and Rose 1985b) and must definitely be borne in mind when choosing and evaluating a method (see Chapter 1).

Thus there were a number of reasons, summarised in Table 3.1, which indicated that the time was ripe to assess the quality of assays of urinary oxalate on a national basis and it was therefore decided to explore the possibility of instigating an external quality assessment (QA) scheme for this analyte.

In September 1984 a letter was inserted in the News Sheet of the Association of Clinical Biochemists in an attempt to ascertain the number of departments that would be interested in participating in a regular QA scheme for urinary oxalate. This produced 34 firm responses and the scheme was initiated with this number of participants in March 1985. This chapter details the findings from the scheme between that date and July 1987.

Organisation of the Quality Assessment Scheme

Details of Participants

The number of regular participants in the scheme had increased from 34 to 46 by July 1987, there having been 20 new members but 8 withdrawals. Some

Table 3.2.   Origin of participants in the quality
assessment scheme

March 1985	
UK	32
Eire	1
Canada	1
Total	34
March 1985–July 1987	
20 additions; 8 withdrawals	
July 1987	
UK	41
Eire	1
Canada	1
USA	1
India	2
Total	46

members are from outside the United Kingdom and the details are shown in Table 3.2.

The questionnaire circulated to all members prior to their participation in the scheme asks for details of the method used plus information concerning workload and assay frequency. This shows that the Sigma kit has continued to be the most popular technique. Details of other methods in use are shown in Table 3.3. Although there is a spectrum of workload represented in the scheme, it is clear that in the majority of laboratories there are fewer than 10 specimens per month and thus the assay is performed rather infrequently (see Figs. 3.1, 3.2).

Table 3.3.   Range of techniques used by participants in the quality assessment scheme

	March 1985	July 1987
Sigma kit	22	32
Precipitation		
Chromotropic acid	6	7
KMnO$_4$ titration	4	1
Atomic absorption	1	0
Oxalate oxidase	0	1
Immobilised oxalate oxidase	1	2
Oxalate decarboxylase/FDH	0	1
High performance liquid chromatography (HPLC)	0	2
Totals	34	46

Preparation and Distribution of Material

The infrequency with which many laboratories perform their urine-oxalate assays indicated that there would be little point in too frequent a circulation since this would lead to "batching" of several distributions. Thus we have chosen to distribute material at six-weekly intervals and there have been eighteen distributions during the period covered by this report. Brief details

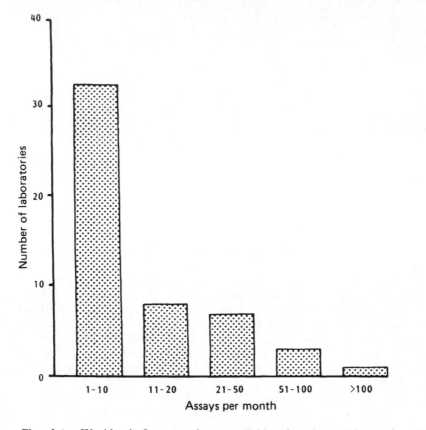

Fig. 3.1. Workload figures, where available, for those laboratories who have participated in the scheme.

of the preparation and nature of the material used during this time are shown in Fig. 3.3. The "stripping" process involves two 30-min adsorptions with alumina (50 g/l of urine) at pH 2.5. One of the problems associated with the preparation of material has been the small but definite loss of oxalate (up to 0.03 mmol/l) that occurs during lyophilisation although this is now minimised by prior adjustment to pH 6.0 and the incorporation of a trace of capryl alcohol. This loss is assessed by analysis of replicate vials before and after lyophilisation and the target value of the spiked pools is corrected for this loss when attempting to assess the accuracy of the returned data. It is accepted that this may not be completely satisfactory although we have satisfied ourselves that the loss is constant from vial to vial (coefficient of variation (CV) consistently less than 4%). The oxalate concentration in the circulated material has ranged from 0.09 mmol/l to 0.95 mmol/l, although an "oxalate-free" preparation has also been distributed. It has been our policy since November 1986 to heat all pooled urine at 60 °C for 90 min prior to lyophilisation, thus minimising any risk from HIV.

Each distribution is accompanied by instructions relating to reconstitution and storage together with the data from the previous distribution. Thus the

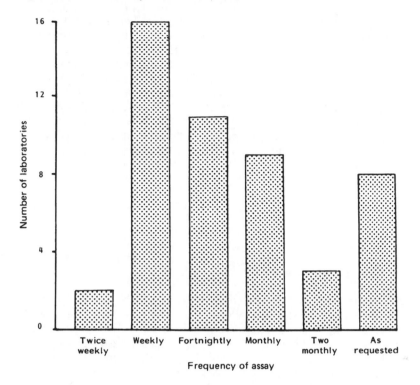

Fig. 3.2. Frequency with which the assay of urinary oxalate is performed in participating laboratories.

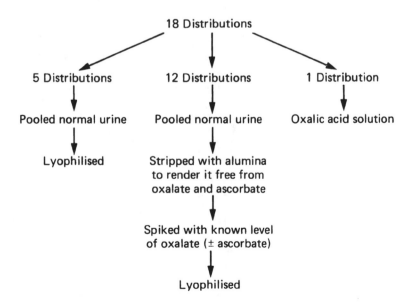

Fig. 3.3. Summary of the material used for the 18 distributions made in the scheme.

results for each participant are listed together with a note of the method used and the pooled data is simply analysed to give an "all-lab" mean, standard deviation and CV. Similar data are also given with the results broken down into the major method groups (ie. Sigma kit with and without removal of ascorbate and chromotropic-acid colorimetry). The result-return rate has varied between 76% and 90% which must be considered satisfactory for such a scheme. Any participant not returning results for 5 consecutive distributions has been withdrawn from the scheme. As with all such external schemes, anonymity between members is preserved, each being identified on paper by means of an allocated code number. The identity of each participant is thus known only to himself and to the organisers.

Results

Selection of Data

When considering the results from the 18 distributions the following should be noted. All the data from distribution 006 has been excluded due to a possible storage problem. In addition the data from 010 (pure oxalic acid) and 018 (zero-oxalate urine) have not been included in any longterm study of imprecision or accuracy. In all studies of accuracy only the data from the 10 accurately spiked pools has been considered. Any calculation based on pooled data for a distribution or method group has been made after the exclusion of any result $> \pm 3$ standard deviations from the mean (SDM).

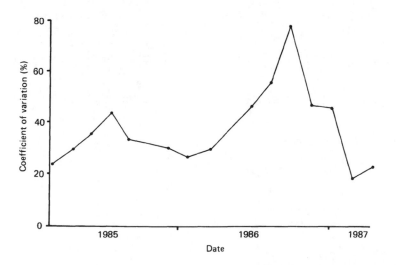

Fig. 3.4. The all-methods CV achieved by participants for each of the distributions across the period of the scheme.

Between-Laboratory Imprecision

The all-methods coefficient of variation achieved by participants for each of the distributions is shown in Fig. 3.4. This has been above 20% on all but one distribution. Figure 3.5 shows data from the two largest method subgroups, ie. the Sigma kit and chromotropic-acid procedures. The Sigma kit data are. then analysed further in Fig. 3.6 on the basis of whether or not an ascorbate-

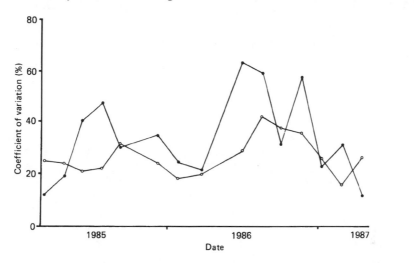

Fig. 3.5. The CV achieved for each of the distributions by the two largest method subgroups (open circles, Sigma kit; filled circles, chromotropic acid).

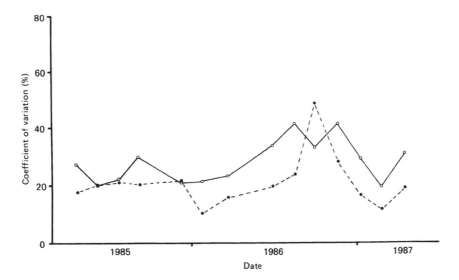

Fig. 3.6. The effect of incorporating an ascorbate removal procedure on the CV figures obtained by the Sigma kit users (open circles, with ascorbate removal; filled circles, without ascorbate removal).

removal procedure was used. Details on this latter aspect were not available for the first distribution in the scheme. Only on one occasion has the group removing ascorbate achieved a lower CV than those not incorporating this stage, and this was with an ascorbate-free urine.

Table 3.4 re-examines the above data in relation to the level of oxalate present (ie. an "imprecision profile"), but using only those results from the accurately spiked pools. The tendency for greater imprecision at the lower concentrations is clearly seen.

Table 3.4. The coefficient of variation (%) figures obtained by the main method groups for the ten spiked pools

Spiked oxalate value (mmol/l)	All	Sigma	Sigma with ascorbate removal	Sigma without ascorbate removal	Chromotropic acid
0.09	56.1	43.0	41.7	23.9	59.9
0.09	78.3	38.3	33.3	48.9	31.9
0.20	47.5	36.3	41.4	28.2	58.0
0.20	46.2	26.2	29.4	16.6	22.8
0.31	44.2	22.4	22.4	21.4	47.6
0.41	18.5	16.1	19.3	11.6	31.8
0.50	33.6	32.1	30.1	20.3	30.3
0.54	30.0	20.0	23.2	15.8	21.7
0.77	23.0	26.4	30.7	18.9	11.9
0.95	26.8'	18.4	21.4	10.3	24.6

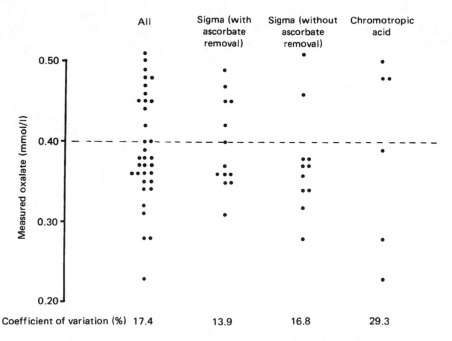

Fig. 3.7. Data obtained from the distribution of a pure oxalic acid solution (0.40 mmol/l).

In an attempt to assess between-laboratory performance in the absence of any potential interference from the urine matrix, and also to check on standardisation procedures, a pure oxalic acid solution was circulated on one occasion (ie. distribution 010). The surprisingly high variation of results obtained from this exercise is shown in Fig. 3.7.

Accuracy

An attempt to assess the accuracy with which the assay is performed was made by examining the results returned on the 10 spiked pools where the "true" oxalate value was known. The all-method mean obtained for each of these distributions is compared with the spiked value in Fig. 3.8. The data from the Sigma kit method are similarly shown in Fig. 3.9, splitting the results into those produced with and without an ascorbate-removal stage. Similar mean values were obtained by both groups unless a high level of ascorbate was present (but note the effect of ascorbate removal on imprecision described elsewhere). The data from those members using a chromotropic-acid procedure are obviously limited but Fig. 3.10 shows the raw data obtained by this technique for each of the spiked pools. The overestimation at low levels of oxalate is particularly obvious.

An attempt was also made to assess the specificity of the various procedures by circulating an oxalate-free urine (as judged by no reaction with oxalate oxidase) although ascorbate was present at a normal level of 0.5 mmol/l. The results of this exercise are shown in Table 3.5. It must be remembered that differences in reporting format can bias this picture. While only 2 laboratories

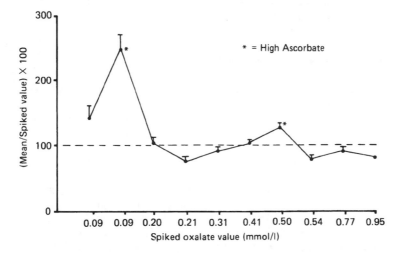

Fig. 3.8. Relationship between the all-method mean and the "true value" for each of the 10 spiked pools. Mean + SEM are indicated at each point.

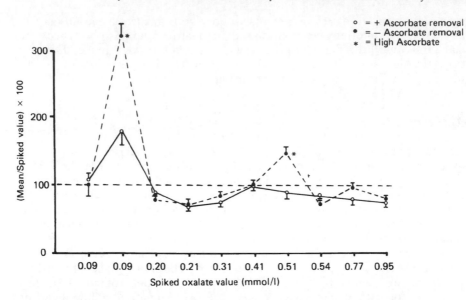

Fig. 3.9. Relationship between the method group mean and the "true value" obtained by Sigma kit users, subdivided on the basis of whether (open circles) or not (filled circles) an ascorbate-removal procedure was used.

reported oxalate levels over 0.1 mmol/l, 19 results (58%) were greater than 0.01 mmol/l. It must be appreciated that the oxalate-stripping procedure may also have removed a number of potential interfering substances and the effect of these will therefore not have been seen.

Ascorbate Interference and its Removal

Two of the spiked urine pools, with oxalate levels respectively 0.5 mmol/l and 0.09 mmol/l, also contained ascorbate at the higher level of 1.0 mmol/l (176 mg/l). Figure 3.11 shows the results from the Sigma kit users for these distributions and illustrates the positive interference from ascorbate in the procedure. For comparative purposes the low-oxalate pool was recirculated, without ascorbate present, and the results of these two distributions for the Sigma kit groupings are seen in Fig. 3.12.

The indication in Fig. 3.6 that the use of an ascorbate-removal step increases imprecision with the Sigma procedure was examined with respect to the two most common ascorbate-removal techniques, these being the pretreatment of urine with ferric chloride or sodium nitrite respectively. Table 3.6 shows the data for these two approaches with respect to the last 5 distributions. Although the data are limited the use of sodium nitrite appears the more satisfactory approach.

Table 3.5. Summary of data obtained from distribution of oxalate-free urine

Method	Number of results	Measured oxalate (mmol/l)				
		"Nil"	≤0.01	0.01<0.05	0.05<0.1	>0.1
All laboratories	33	3	11	15	2	2
Sigma kit	23	2	10	9	0	2
Sigma kit with ascorbate removal	13	2	5	5	0	1
Sigma kit without ascorbate removal	10	0	5	4	0	1
Chromotropic acid	5	0	0	3	2	0

Table 3.6. Ferric chloride and sodium nitrite-based procedures compared as methods for ascorbate removal

Distribution number	Oxalate (mmol/l)	Ascorbate (mmol/l)	No ascorbate removal			Ferric chloride			Sodium nitrite		
			n	Oxalate mmol/l	CV%	n	Oxalate mmol/l	CV%	n	Oxalate mmol/l	CV%
013	0.09	Nil	9	0.090	48.9	11	0.094	38.3	4	0.105	23.4
014	0.21	0.10	11	0.149	28.2	10	0.151	51.0	4	0.133	7.5
015	0.20	0.50	11	0.157	16.6	9	0.164	39.0	4	0.203	6.4
016	0.41	0.50	12	0.414	11.6	9	0.391	25.7	5	0.426	6.3
017	0.77	0.50	12	0.766	18.9	8	0.520	31.3	5	0.832	10.0

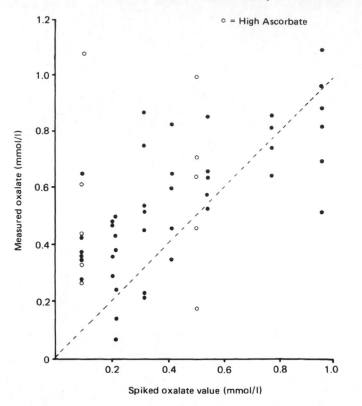

Fig. 3.10. Raw data obtained on the spiked pools by those members using a chromotropic acid procedure. The line of equivalence is shown, and open circles indicate high ascorbic acid content.

Other Observations on Selected Methods

The data from the members using a chromotropic-acid procedure have already been referred to, but two other interesting observations emerge from a study of the results from the less commonly used techniques. Of the 4 original participants who used a precipitation/potassium-permanganate titration procedure only one remains, the others having either withdrawn or changed to the Sigma kit method. However this remaining user (Lab X) has consistently achieved a satisfactory performance during the period of the scheme, in contrast to the marked negative bias shown by virtually all other results produced using this method (see Fig. 3.13).

Also of interest are the results generated by the two laboratories using a fully automated continuous-flow procedure involving immobilised oxalate oxidase. The satisfactory performance achieved by this approach is illustrated in Fig. 3.14. Of the other techniques in use by participants none has generated sufficient results to allow any comment to be made.

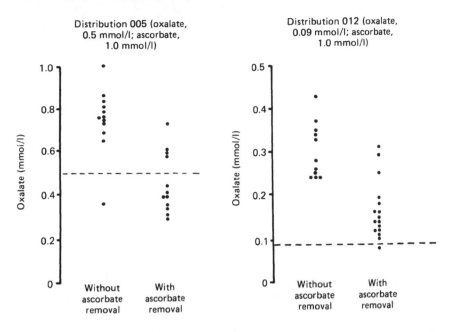

Fig. 3.11. Data from the two high-ascorbate pools showing the effect of using an ascorbate-removal procedure on the results obtained using the Sigma kit.

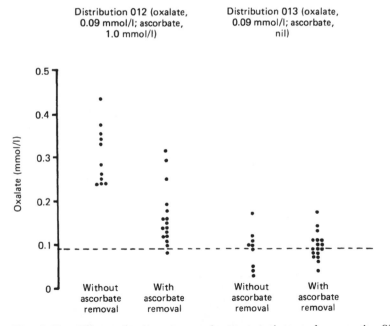

Fig. 3.12. Effect of using an ascorbate-removal procedure on the Sigma-kit data obtained from a low-oxalate pool in the presence and absence of ascorbate.

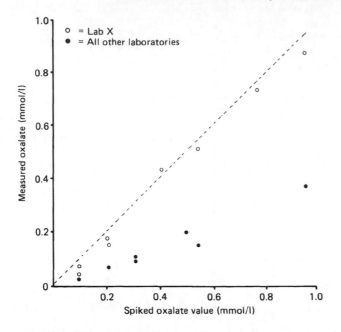

Fig. 3.13. Results obtained at various levels from those participants using a precipitation/ potassium permanganate titration approach. Open circles, Lab X; filled circles, all other laboratories.

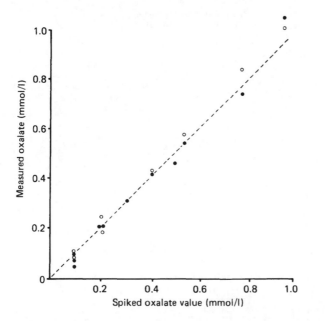

Fig. 3.14. Results obtained at various levels by the two participants using a continuous flow/immobilised oxalate oxidase procedure.

Discussion

It must be realised that the small number of participants in this scheme, combined with a 6-weekly distribution programme, has meant that the amount of data generated during the first 30 months of operation has inevitably been limited, allowing only preliminary observations to be possible. In addition it is accepted that the practice of using oxalate-free urine pools and then spiking to known levels of oxalate prior to lyophilisation, as has been done for several distributions, may not exactly mimic the effect of urine in some assays since other potentially interfering substances may be removed by the stripping procedure. We feel, however, that a number of significant points have already emerged concerning the quality of performance currently being achieved in urine oxalate assays.

The agreement between laboratories in the scheme as judged by the all-method CV was disturbingly poor, values ranging from 18.5% to 78.3% over the period studied, with no real sign of improvement. Considering individual methods, some particularly high CV values occurred in the group using the chromotropic-acid technique although based on an average of only 6 results per distribution. The CV values within the Sigma kit users, the largest single method group, ranged from 16.1% to 43.0% and further examination of these data showed that, on all but one distribution, the imprecision was greater within the subgroup of laboratories who were incorporating an ascorbate-removal step in the procedure, this important aspect being discussed further below. Generally, imprecision was greatest at the lowest oxalate concentration (ie. 0.09 mmol/l) although it must be appreciated that many patients with marginal hyperoxaluria will have urine concentrations around this level if their daily urine volume is high, as is usually the case with patients attending stone clinics and instructed to maintain a high fluid intake. Thus precision in this region is highly desirable. Only marginal improvement in the agreement between laboratories was afforded when lyophilised urine was replaced by a pure oxalic acid solution, thus removing any interference from the urine matrix. For example, the CV obtained by the Sigma kit group for this material was 15.3% (compared with 16.1% for urine at a similar concentration) and presumably all these users are calibrating against the same kit standard.

Perhaps one of the most interesting findings to emerge from the scheme so far relates to the problem of interference by ascorbic acid. The two questions that need to be answered here are firstly, does ascorbate interfere to a significant degree in the various procedures and secondly, what is the most reliable way of removing this interference? This area has been the subject of much recent debate, and numerous personal communications with participants have confirmed that guidance is badly needed. The presence of ascorbic acid at a concentration of 1.0 mmol/l in two distributions seemed to have little extra effect on the very wide scatter of results yielded by the chromotropic-acid method, but the Sigma kit data from these pools were more meaningful. On both occasions a clear positive bias was observed within the group not employing an ascorbate removal step, mean (±1 SD) values of 0.75 (±0.15) mmol/l and 0.29 (±0.07) mmol/l being obtained from pools spiked to oxalate levels of 0.50 and 0.09 mmol/l respectively. In addition at the lower oxalate level a positive bias was still seen (ie. 0.16 $\pm$ 0.07 mmol/l) within the group using ascorbate-removal techniques, suggesting that these were not completely efficient. This seemed to be confirmed by the results obtained from circulating this same low-oxalate pool (0.09 mmol/l) but with a zero-ascorbate content, similar results now being obtained whether ascorbate removal was or was not attempted, ie. 0.096 ± 0.032 mmol/l and 0.090 ± 0.044 mmol/l respectively. Thus although there is the potential for both positive and negative interference by ascorbate in the Sigma procedure, the net effect seems to be the production of falsely elevated results, indicating that some removal procedure is desirable. In this context it should be noted that the above effects were produced with an ascorbate level of 1.0 mmol/l, a concentration easily exceeded by consuming ascorbate-rich food and drinks and certainly by megadose ascorbic-acid supplementation as practised by some individuals for dubious health reasons.

With respect to the choice of a procedure for the prior removal of ascorbate the data from the scheme suggests that pretreatment of urine with ferric chloride is unsatisfactory since it increases the imprecision of the Sigma procedure. However, the limited amounts of data from those laboratories using pretreatment with sodium nitrite is encouraging, and is in agreement with recently published work relating to the problems of the Sigma kit procedure (Barlow 1987; Kasidas and Rose 1987). In addition, at least two participants have improved their performance in the scheme by changing from a ferric chloride to a sodium nitrite based procedure.

The results from the single distribution of "oxalate-free" urine showed that 4 laboratories, 2 using the Sigma kit and 2 using a chromotropic-acid technique, reported levels of greater than 0.05 mmol/l which is probably unacceptable. It would seem desirable to have non-specific interference at a level of less than 0.01 mmol/l if significant errors are not to be made in situations where urine volumes are high. This was apparently achieved by only 14 of the 33 returning laboratories.

In view of the generally poor performance seen during the period of operation of the scheme it is difficult to make firm conclusions concerning the choice of method for routine assay of urinary oxalate. However, it would seem fair to conclude that those methods based on oxalate precipitation followed by potassium permanganate titration or by chromotropic acid colorimetry cannot be recommended. With one notable exception the former approach yielded low results while the chromotropic-acid technique tended to overestimate and, furthermore, the within- and between-laboratory imprecision was high. The Sigma kit procedure, incorporating a sodium nitrite based ascorbate removal step, may prove to be the most satisfactory approach for relatively low workloads, although the colorimetric stage can be automated on various discrete and centrifugal analysers. For high workloads, the impressive performance of the two participants using a continuous-flow technique based on immobilised oxalate oxidase is worthy of note. Indeed it may well be that one approach to improving the quality of performance of this assay could be to centralise the workload in those departments able to set up more sophisticated procedures, bearing in mind the infrequency with which several participants said they performed the estimation. In this context it will be interesting to see how the high performance liquid chromatography (HPLC) procedures perform as they are established in the scheme.

Conclusion

In summary, it would appear that our feelings concerning the need to examine the performance of urinary oxalate assays have been more than justified. The early findings are in general agreement with other studies (Zerwekh et al. 1983; Shepherd and Penberthy 1987) and indicate that in many laboratories the quality of results does not meet the clinical need, as it does not allow for the detection and monitoring of the mild hyperoxaluria now thought to be so important in the calcium oxalate stone former. It is to be anticipated that as the scheme enlarges (and it should be noted that the organisers know of several laboratories performing the assay who are not yet members) we will be able to define better the problems relating to the various techniques and their modifications. This should allow firmer conclusions to be made concerning choice of methods and hopefully lead to much needed improvement in the performance of this assay.

Acknowledgements. I am grateful to Dr. G. P. Kasidas for the regular preparation of the lyophilised urine samples. Thanks are also due to the secretarial staff of the Biochemistry Department.

References

Antonacci A, Colussi G, De Ferrari ME et al. (1985) Calcium oxalate urine supersaturation in calcium stone formers: hypercalciuria versus hyperoxaluria. In: Schwille PO, Smith LH, Robertson WG, Vahlensieck W (eds) Urolithiasis and related clinical research. Plenum Press, New York, pp 283–286

Baggio B, Gambaro G, Favaro S et al. (1983) Prevalence of hyperoxaluria in idiopathic calcium oxalate kidney stone disease. Nephron 35: 11–14

Bais R, Potezny N, Edwards JB et al. (1980) Oxalate determination by immobilised oxalate oxidase in a continuous flow system. Anal Chem 52: 508–511

Barlow IM (1987) Obviating interferences in the assay of urinary oxalate. Clin Chem 33: 855–858

Buttery JE, Ludvigsen N, Braiotta EA et al. (1983) Determination of urinary oxalate with commercially available oxalate oxidase. Clin Chem 29: 700–702

Chalmers AH, Cowley DM (1984) Urinary oxalate by rate analysis compared with gas chromatographic and centrifugal analyser methods. Clin Chem 30: 1891–1892

Costello J, Hatch M, Bourke E (1976) An enzymic method for the spectrophotometric determination of oxalic acid. J Lab Clin Med 87: 903–908

Hallson PC, Rose GA (1974) A simplified and rapid enzymatic method for the determination of urinary oxalate. Clin Chim Acta 55: 29–39

Hodgkinson A (1977) Oxalic acid in biology and medicine. Academic Press, New York, pp 62–103

Jaeger P, Portmann L, Jacquet A et al. (1985) Influence of the calcium content of the diet on the incidence of mild hyperoxaluria in idiopathic renal stone formers. Am J Nephrol 5: 40–44

Kasidas GP, Rose GA (1985a) Continuous flow assay for urinary oxalate using immobilised oxalate oxidase. Ann Clin Biochem 22: 412–419

Kasidas GP, Rose GA (1985b) Spontaneous in vitro generation of oxalate from L-ascorbate in some assays for urinary oxalate and its prevention. In: Schwille PO, Smith LH, Robertson WG, Vahlensieck W (eds) Urolithiasis and related clinical research. Plenum Press, New York, pp 653–656

Kasidas GP, Rose GA (1987) Removal of ascorbate from urine prior to assaying with a commercial oxalate kit. Clin Chim Acta 164: 215–221

Kohlbecker G (1981) Direct spectrophotometric determination of serum and urinary oxalate with oxalate oxidase. J Clin Chem Clin Biochem 19: 1103–1106

Laker MF, Hofmann AF, Meeuse BJD (1980) Spectrophotometric determination of urinary oxalate with oxalate oxidase prepared from moss. Clin Chem 26: 827–830

Mahle CJ, Menon M (1982) Determination of urinary oxalate by ion chromatography: preliminary observation. J Urol 127: 159–162

Mazzachi BC, Teubner JK, Ryall RL (1984) Factors affecting measurement of urinary oxalate. Clin Chem 30: 1339–1343

McWhinney BC, Cowley DM, Chalmers AH (1986) Simplified liquid chromatographic method for measuring urinary oxalate. J Chromatogr. 383: 137–141

Obzansky DM, Richardson KE (1983) Quantification of urinary oxalate with oxalate oxidase from beet stems. Clin Chem 29: 1815–1819

Robertson WG, Nordin BEC (1969) Activity products in urine. In: Hodgkinson A, Nordin BEC (eds) Renal Stone Research Symposium. Churchill, London, pp 221–232

Robertson WG, Peacock M (1980) The cause of idiopathic calcium stone disease: hypercalciuria or hyperoxaluria: Nephron 26? 105–110

Robertson WG, Rutherford A (1980) Aspects of the analysis of oxalate in urine – a review. Scand J Urol Nephrol [Supp] 53: 85–95

Robertson WG, Peacock M, Heyburn PJ et al. (1979) The significance of mild hyperoxaluria in calcium stone formation. In: Rose GA, Robertson WG, Watts RWE (eds) Oxalate in biochemistry and clinical pathology. Wellcome Foundation, London, pp 173–180

Shepherd MDS, Penberthy LA (1987) Performance of quantitative urine analysis in Australasia critically assessed. Clin Chem 33: 792–795

Zerwekh JE, Drake E, Gregory J et al. (1983) Assay of urinary oxalate: six methodologies compared. Clin Chem 29: 1977–1980

Chapter 4

Assaying of Oxalate in Plasma

G. P. Kasidas

Introduction

Until recently the reliable measurement of oxalate in plasma presented considerable problems. Primarily, this was due to the extremely low level of the anion in plasma. The range of oxalate levels reported in normal plasma varied enormously (see Table 4.1). Such wide variations in the levels of oxalate in plasma probably arose from the difficulty in developing highly sensitive and specific assays to detect the low level of oxalate in plasma and also from the rapid in vitro oxalogenesis from glyoxylate (Akcay and Rose 1980), ascorbate (Kasidas and Rose 1986) and other precursors (France et al. 1985) during the storage and processing of samples. The problem of oxalogenesis from ascorbate in urine was discussed in Chapter 2. In plasma, because of the neutral/near-alkaline pH environment and the relatively higher concentration ratio of ascorbate to oxalate which is normal in this matrix, oxalogenesis from ascorbate may pose even greater problems (see below). In vivo radioisotope assays (Williams et al. 1971; Pinto et al. 1974; Hodgkinson and Wilkinson 1974; Constable et al. 1979; Prenen et al. 1982) have consistently yielded plasma oxalate levels in normal subjects of 0.4–2.8 μmol/l whereas the earlier in vitro assays (see Table 4.1) produced values ranging from 30 to 300 μmol/l. The gap between the two approaches for determining plasma oxalate has narrowed considerably as a result of recognition and subsequent correction of the problems arising from spontaneous in vitro oxalogenesis and the development of highly specific and sensitive assays.

The inability to measure plasma oxalate accurately has also left a gap in the understanding of the normal handling of oxalate by the intact human kidney.

Plasma Oxalate Determination

Separation of Oxalate from Plasma prior to its Measurement

Plasma contains proteins and other macromolecules and because these can interfere in assay procedures they usually are removed prior, and in addition, to the other separation steps already discussed in Chapter 2. Acetone (Nuret and Offner 1978), acidification and heat treatment (Zarembski and Hodgkinson 1965; Boer et al. 1984; Akcay and Rose 1980) and ultrafiltration (Bennett et al. 1979; Hatch et al. 1977; Ichiyama et al. 1985; Kasidas and Rose 1986) have all been used to deproteinise plasma prior to the determination of oxalate. Almost complete recoveries of ^{14}C-labelled oxalate (<95%) have been demonstrated in the ultrafiltration (Ichiyama et al. 1985), acidification and heat-treatment (Boer et al. 1984) procedures. However, in some enzymatic assays (Crawhall and Watts 1961; Samsoondar et al. 1983) the measurement of oxalate was attempted in plasma without removal of the macromolecules.

In vitro Assays for Plasma Oxalate

Chemical Techniques

As early as the fourth decade of this century measurement of oxalate in plasma by chemical assays had been attempted (Merz and Maugeri 1931; Suzuki 1934). These investigators separated the oxalate from deproteinised plasma as calcium oxalate prior to its measurement by titration with potassium permanganate. Values of plasma oxalate obtained by them ranged between 222 and 444 μmol/l. Subsequently these levels were disputed by Thomsen (1935) who questioned the specificity of the assays. He pointed out that calcium oxalate which had been precipitated was contaminated with other substances which, during the assay, yielded erroneously higher values of plasma oxalate. He concluded from his study that the oxalate level in plasma was less than 111 μmol/l.

Attempts were made (Flaschenträger and Muller 1938; Barber and Gallimore 1940; Barrett 1943) to improve the specificity of titrimetric assays by elaborate and tedious separation procedures. These involved esterification followed by distillation, hydrolysis and final precipitation of the oxalate prior to its titration with potassium permanganate. These procedures were not only tedious but also contained steps in which oxalogenesis can occur. However, lower values of plasma oxalate (range 22–89 μmol/l) were reported by these investigators.

Pernet and Pernet (1965) moved away from conventional titrimetric procedures to develop a colorimetric procedure. They precipitated the oxalate as its lead salt and chemically reduced it to glyoxylate before colorimetric estimation with phenylhydrazine and ferricyanide. A mean value of plasma oxalate of 25 μmol/l was found by these workers. In another colorimetric procedure, Krugers Dagneaux et al. (1976) separated the oxalate from other interfering constituents of plasma by anion-exchange chromatography and reduced it to glycollate before colorimetric estimation with chromotropic acid.

Levels ranging between 13 and 28 μmol/l for plasma oxalate were found using this technique.

Assays based on fluorimetry have been reported (Zarembski and Hodgkinson 1965; Endo 1969). Zarembski and Hodgkinson (1965) separated the oxalate from interferants by first extracting the oxalate in plasma with the solvent tri-n-butylphosphate (see Chap. 2) followed by coprecipitation with calcium sulphate. They chemically reduced the precipitated oxalate to glyoxylate and then determined this compound fluorimetrically after reacting it with resorcinol. The reduction of oxalate to glyoxylate by zinc and hydrochloric acid is difficult to control and errors could have been introduced in this step.

Most of the chemical assays for oxalate in plasma have been tedious, time-consuming and often were not specific for oxalate. Moreover, they consisted of steps which possibly could have promoted oxalogenesis (see above and below).

Physical Techniques

In the search for a simpler and more reliable assay some investigators have turned to physical techniques for measuring plasma oxalate. Gas chromatography (Chambers and Russell 1973; Nuret and Offner, 1978; Gelot et al. 1980; Wolthers and Hayer 1982; Lopez et al. 1985) and high performance liquid chromatographic (HPLC) (Jerez 1986; Ramsay and Reed 1984) procedures have been described. Gas-chromatographic (GC) procedures invariably required preliminary separation of the oxalate from plasma constituents followed by its derivatisation to volatile esters before injection in gas-chromatographic columns for separation and subsequent quantitation. The earlier GC procedures (Chambers and Russell 1973; Nuret and Offner 1978; Gelot et al. 1980) suffered from drawbacks such as poor resolution, inadequate sensitivity and derivatisation under conditions at which oxalogenesis can occur (Tocco et al. 1979). Values of plasma oxalate obtained by these assays were much higher than those obtained by the in vivo radioisotope assays.

Wolthers and Hayer (1982) and Lopez et al. (1985) described capillary gas-chromatographic procedures for plasma oxalate measurement. They claimed that, unlike the conventional packed columns, capillary gas-chromatographic columns can produce better resolutions and sensitivity. They carried out their derivatisation under acidic conditions and they also took other special precautions with sample collection to minimise oxalogenesis and as a result obtained values of plasma oxalate in the same order of magnitude (1–6 μmol/1) (see Table 4.1) as radioisotopic assays.

Liquid-chromatographic separation and quantitation of oxalate in an HPLC-assay system at ambient temperature was proposed recently by Jerez (1986). An elaborate preliminary separation procedure involved precipitating oxalate from plasma ultrafiltrate, dissolving the precipitated oxalate, solvent extraction of the dissolved oxalate with diethyl-ether, evaporating to total dryness and a second dissolution in a weak phosphoric acid mobile phase before HPLC separation and quantitation. Not surprisingly relatively higher values for plasma oxalate (32 ± 10 μmol/l) were obtained and this could be attributed to the long time at which the sample is kept at conditions of neutral/near-alkaline pH which favours oxalogenesis from ascorbate (Rose 1985).

Table 4.1. Reported levels of plasma oxalate in normal subjects

Authors	Year	Plasma oxalate (μmol/l)		Number of subjects	Technique
		Mean	Range		
In vitro chemical/enzymatic/physical assays					
Merz and Maugeri	1931	—	222–444	—	Titrimetric
Pernet and Pernet	1965	25	19–32	—	Colorimetric
Krugers Dagneaux et al.	1976	—	13–28	20	Colorimetric
Zarembski and Hodgkinson	1965	—	15–31	15	Fluorimetric
Endo	1969	28	17–53	—	Fluorimetric
Hatch et al.	1977	20	8–52	40	Enzymatic
Knowles and Hodgkinson	1972	13	9–16	20	Enzymatic
Akcay and Rose	1980	2.3	0–5.4	20	Enzymatic
Boer et al.	1984	3.3	1.2–6.4	24	Enzymatic
Ichiyama et al.	1985	3.5	—	8	Enzymatic
Samsoondar et al.	1983	161	110–210	15	Enzymatic
Kohlbecker	1981	33	17–45	10	Enzymatic
Kasidas and Rose	1986	2.03	1.3–3.1	21	Enzymatic
Parkinson et al.	1985	—	0.8–1.5	35	Enzymatic
France et al.	1985	13	4–22	36	Enzymatic
Bennett et al.	1979	10	5–17	15	Radioenzymatic
Cole et al.	1984	4.0	—	—	Radioenzymatic
Chambers and Russell	1973	161	62–260	20	Gas chromatographic
Gelot et al.	1980	20	9–41	40	Gas chromatographic
Nuret and Offner	1978	16	12–20	20	Gas chromatographic
Wolthers and Hayer	1982	2.8	1–5	22	Capillary gas chromatographic
Lopez et al.	1985	3.9	2.7–6.0	16	Capillary gas chromatographic
Jerez	1986	32	18–50	16	HPLC
In vivo radioisotopic assays					
Williams et al.	1971	2.0	1.4–2.3	6	Isotopic
Pinto et al.	1974	1.4	0.4–2.8	6	Isotopic
Hodgkinson and Wilkinson	1974	1.4	1.3–1.6	3	Isotopic
Constable et al.	1979	1.3	0.8–2.1	8	Isotopic
Prenen et al.	1982	1.4	1.0–1.8	10	Isotopic

Enzymatic Techniques

The specificity and relatively mild conditions of assay offered by the use of enzymes makes them attractive for their application to the measurement of plasma oxalate. Numerous enzymatic procedures (Table 4.1) using either of the two oxalate-decomposing enzymes (oxalate decarboxylase EC4.1.1.3 or oxalate oxidase EC1.2.3.4) (see Chapter 2) for plasma oxalate have been reported.

Crawhall and Watts (1961) added oxalate decarboxylase to neat plasma and determined the carbon dioxide formed by Warburg manometry. They concluded that the level of oxalate in normal plasma is less than 88 μmol/l, below the lower limit of sensitivity of the assay. Some years later Knowles and Hodgkinson (1972), using the same enzyme, improved the sensitivity by determining the carbon dioxide colorimetrically after trapping it in a weakly alkaline solution of phenolphthalein. Although this assay was automated, it was still laborious and great care had to be taken to prevent aerial contamination of the alkaline phenolphthalein by carbon dioxide.

Akcay and Rose (1980) also determined plasma oxalate with oxalate decarboxylase but by measuring the change in pH of a weak alkaline buffer arising from the trapping of carbon dioxide evolved in the reaction. These investigators admitted that the method was highly demanding, tedious and required large volumes of blood which made it unsuitable for routine use. However, they did demonstrate that oxalogenesis from glyoxylate can occur in blood.

Bennett et al. (1979) described a radioenzymatic isotope-dilution assay using oxalate decarboxylase. They determined the level of plasma oxalate by adding ^{14}C-labelled oxalate and measuring the relative radioactivities of the carbon dioxide evolved in labelled and unlabelled samples. Values of plasma oxalate ranging between 5 and 17 μmol/l were obtained. Cole et al. (1984) subsequently refined sample collection procedures and took measures to prevent oxalogenesis before reapplying the above assay for the determination of normal levels of plasma oxalate. A lower mean value of 4.0 μmol/l for plasma oxalate was thus found.

Recently, with the availability of oxalate oxidase, several assays (Samsoondar et al. 1983; Ichiyama et al. 1985; Sugiura et al. 1980; France et al. 1985; Boer et al. 1984; Kasidas and Rose 1986) for plasma oxalate using this enzyme have been described. Boer et al. (1984) utilised the fact that two molecules of carbon dioxide are liberated from the enzymatic oxidation of oxalate compared to the single molecule liberated from the decarboxylation of oxalate. They measured the change in pH of a weak alkaline buffer arising from the trapping of the evolved carbon dioxide. Special precautions with sample-collection and use of acid conditions of assay yielded normal plasma oxalate levels close to those obtained by the in vivo radioisotopic assays (range 1.2–6.4 μmol/l).

With the limitations imposed on the detection and quantitation of the evolved carbon dioxide some investigators turned to measuring the formate produced in the enzymatic decarboxylation of oxalate. Hatch et al. (1977) reported a double-enzyme assay system for plasma oxalate. They used oxalate decarboxylase in a linked reaction with NAD$^+$-requiring formate dehydrogenase (EC1.2.12) (see Chap. 2). Neutral/near-alkaline conditions which favour conversion of ascorbate to oxalate were used in their separation and assay procedures and this led to the plasma oxalate level being erroneously overestimated. Values of plasma oxalate ranging between 8 and 52 μmol/l were reported.

More recently, Parkinson et al. (1985) reported a multi-enzyme bioluminescent assay system for determining plasma oxalate. Like the previous investigators (Hatch et al. 1977) they used oxalate decarboxylase at pH 3 to produce formate and then enzymatically dehydrogenated it with formate dehydrogenase to produce NADH which they determined by a bioluminescent reaction with bacterial luciferase. Values of plasma oxalate ranging between 0.8 and 1.5 μmol/l were reported by these investigators.

The enzyme oxalate oxidase also yields hydrogen peroxide as one of its products. Numerous proven peroxidase-mediated colorimetric procedures are available for the measurement of hydrogen peroxide (Carr and Bowers 1980). Several methods (Sugiura et al. 1980; Ichiyama et al. 1985; France et al. 1985; Kasidas and Rose 1986; Rehmert et al. 1983; Samsoondar et al. 1983) for determining plasma oxalate using oxalate oxidase coupled with peroxidase-mediated colorimetric techniques have been reported. A wide range of values (1.2–210 μmol/l) for plasma oxalate has been found by these assays. Higher values seem to be associated with lack of precautions to minimise oxalogenesis.

It is now becoming generally accepted that true plasma oxalate levels are in the same order of magnitude as those determined by the in vivo radioisotopic assays. By using highly specific and sensitive assays and taking extra care to prevent oxalogenesis the in vivo isotopic and in vitro chemical/physical/enzymatic approaches yield similar levels for normal plasma oxalate.

In Vivo Radioisotopic Assays

Estimation of plasma oxalate by indirect in vivo assays (see Table 4.1) has been proposed. These assays are based on the principle of infusing or injecting intravenously ^{14}C-labelled oxalate and allowing the label to equilibrate in body compartments. At equilibrium it is assumed that the specific radioactivities of plasma and urine are constant and equal thus:

Plasma radioactivity/Plasma oxalate = Urine radioactivity/Urine oxalate

The radioactivities of plasma and urine are determined and urine oxalate concentrations measured by isotope dilution (Williams et al. 1971), colorimetry (Prenen et al. 1982; Hodgkinson and Wilkinson 1974) or enzymatic (Constable et al. 1979) methods. By substitution of three of the four variables in the above equation, the plasma oxalate is calculated.

Williams et al. (1971) continuously infused ^{14}C-labelled oxalate and measured plasma and urine radioactivities. Their technique for measuring urinary oxalate also relied on carbon-14 which made them assume a value for this in order to calculate the corresponding plasma oxalate concentration. Hodgkinson and Wilkinson (1974) gave single doses of ^{14}C-labelled oxalate, counted plasma and urine radioactivities and measured urinary oxalate by a colorimetric technique. Continuously falling radioactivity raises theoretical objections and this may explain their inconsistent results. However, Prenen et al. (1982) subsequently found that constant infusion when simultaneously compared with single-bolus injection techniques produced identical levels of oxalate in plasma. Constable et al. (1979) estimated plasma oxalate by a continuous infusion in vivo radioisotopic technique. Urine oxalate was measured enzymatically with oxalate decarboxylase by the assay of Hallson and Rose (1974). At high levels of plasma creatinine (≤500 μmol/l) they were also able to measure the oxalate in plasma directly by the enzymatic urine oxalate assay. When the in vitro enzymatic and in vivo radioisotopic assays had been applied simultaneously at these high levels of plasma oxalate a good agreement between them was found (Constable et al. 1979).

Possible health hazards associated with the use of radiolabelled tracers in healthy subjects make in vivo radioisotopic assays unsuitable for routine use.

Nevertheless they served in a research environment in establishing standards for the development of assays more suitable for routine use and through them led to the postulation and finding of oxalogenesis during storage and collection of plasma samples.

Assaying of Plasma Oxalate at St Peter's Hospitals (London)

At St Peter's Hospitals a continuous flow assay (Kasidas and Rose 1986) with immobilised oxalate oxidase is used to measure the level of oxalate in plasma. A summary of this assay will be given here.

Blood is obtained by venepuncture and immediately heparinised, centrifuged (1614 g for 10 min) at 4 °C, the plasma separated and frozen within 1 h and kept frozen until it can be processed further. Plasma (5 ml) is acidified to pH 3.6–4.2 by addition of 25 μl concentrated hydrochloric acid and then ultrafiltered through an Amicon Centriflo membrane, Type CF25, with a molecular-weight cut-off of 25 000 daltons. The ultrafiltrate is treated with sodium nitrite (5 mmol/l) and then diluted eight times with 0.05м-citrate buffer at pH 3.5 and analysed in an AutoAnalyser system using a nylon coil bearing immobilised oxalate oxidase. The oxalate generates hydrogen peroxide which in turn generates a colour with peroxidase, 3-methyl-2-benzothiazolanone hydrazone (MBTH) and 3-dimethylamino-benzoic acid (DMAB).

The recovery of added oxalate from plasma samples from ten normal subjects and from patients with chronic renal failure by this method was $96 \pm 8.82\%$ (mean $\pm$ SD). Figure 4.1 shows that when five samples of plasma were allowed to stand at room temperature for 18 h the measured levels of oxalate rose steadily from starting values of 2–3 μmol/l to 7–15 μmol/l. Similar rises were found when whole blood or ultrafiltrates (without acidification) were allowed to stand at room temperature: oxalogenesis was therefore not enzyme-dependent. Figure 4.2 shows that oxalogenesis was prevented by addition of sodium nitrite or acidification or both. This method has the following advantages:

Sample preparation is not too elaborate

Oxalogenesis from ascorbate is controlled

Speed, cost-effectiveness (cost of reagent and ultrafiltration membrane is £0.30 per assay) and improved performance characteristics are derived from automation and reusability of the enzyme reactor

The same continuous flow equipment and enzyme reactor are used for urinary oxalate estimations

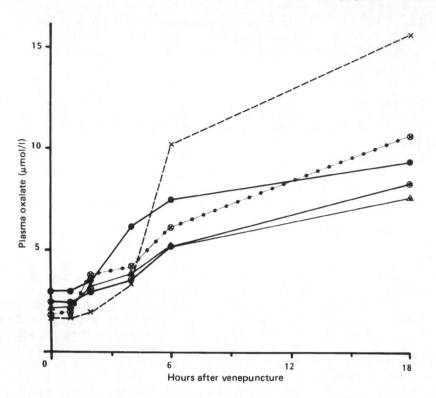

Fig. 4.1. Spontaneous generation of oxalate at physiological pH in 5 plasma samples left standing at room temperature. Reproduced with permission from Kasidas GP and Rose GA (1986) Clin Chim Acta 154: 49–58.

Plasma Oxalate Levels in Normal and Some Pathological States. Clinical Usefulness of Plasma Oxalate

Normal Subjects

The distribution of plasma oxalate levels in 21 healthy adults is shown in Fig. 4.3. These subjects were kept on unrestricted dietary regimens and invariably samples were taken during the mid-morning period.

Patients with Renal Failure with or without Primary Hyperoxaluria

Oxalate is a useless end-product of metabolism and is removed from the blood mainly by the kidneys. Several studies (Constable et al. 1979; Zarembski et al. 1966; Kasidas and Rose 1986; Boer et al. 1984; Ahmad and Hatch

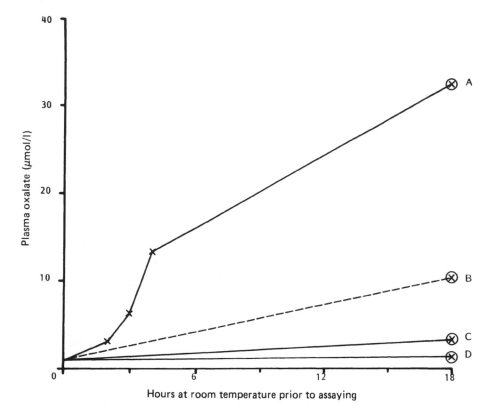

Fig. 4.2. Conversion of L-ascorbate to oxalate in a pooled plasma sample and its prevention by sodium nitrite. *A* A portion of the pooled plasma spiked with L-ascorbic acid (570 μmol/l), *B* A portion without added L-ascorbic acid; *C*, with added L-ascorbic acid (570 μmol/l) and sodium nitrite (5 mmol/l); and *D* without added ascorbic acid but with added $NaNO_2$. Reproduced with permission from Kasidas GP and Rose GA (1986) Clin Chim Acta 154: 49–58.

1985) have demonstrated that oxalate accumulates proportionately with the deterioration of renal function. Figure 4.4 shows a linear relationship ($y = 26.02x - 0.89$; R = 0.92 p < 0.001) between oxalate and creatinine in plasma of 27 patients with chronic renal failure. Some of the oxalate thus retained by the impaired renal function can be removed by haemodialysis (Rehmert et al. 1983; Ramsey and Reed 1984; Ahmad and Hatch 1985; Boer et al. 1984; Borland et al. 1987; O'Reagan et al. 1979) but levels rise again soon after dialysis of the patient is stopped (Zarembski et al. 1966; Borland et al. 1987). Oxalosis can occur in chronic renal failure without concomitant hyperoxaluria (Salyer and Keren 1973; Macaluso and Berg 1959): hence it would be difficult to prove the presence of hyperoxaluria in anuric patients. However by measuring the plasma oxalate/creatinine concentration ratio it is possible to obtain an indication of the production rate of oxalate. Kasidas and Rose (1987) used the oxalate/creatinine concentration ratio to prove the presence of hyperoxaluria when renal failure develops as a result of it. Figure 4.5 shows the relationships of oxalate with creatinine in plasma of patients with renal

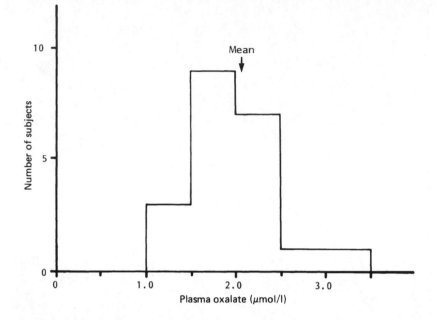

Fig. 4.3. Distribution of plasma oxalate in 21 healthy adult subjects. (Mean, 2.03; SD, 0.52 μmol/l). Reproduced with permission from Kasidas GP and Rose GA (1986) Clin Chim Acta 154: 49–58.

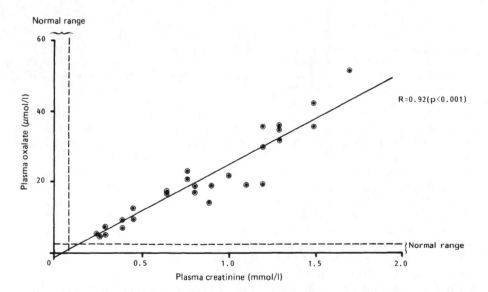

Fig. 4.4. Relationship between oxalate and creatinine levels in plasma in chronic renal failure. Line of regression ($y = 26x - 0.89$) is shown. Reproduced with permission from Kasidas GP and Rose GA (1986) Clin Chim Acta 154: 49–58.

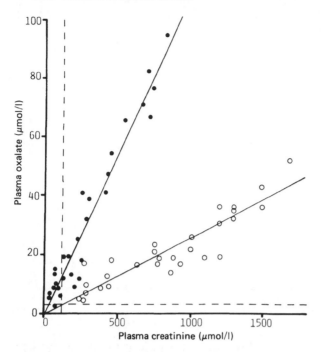

Fig. 4.5. Relationship between oxalate and creatinine levels in plasma in patients with primary hyperoxaluria (filled circles) and in patients with renal failure from other causes (open circles). Upper limits of the normal ranges are shown by the broken lines.

failure, with or without primary hyperoxaluria. High plasma concentration ratios of oxalate/creatinine (> 0.038) are associated with primary hyperoxaluria with chronic renal failure and values below this are associated with chronic renal failure only.

Monitoring of plasma oxalate levels is also useful to indicate whether they are sufficiently high to be causing oxalosis particularly when patients are to receive kidney grafts. Kasidas and Rose (1987) suggested that plasma oxalate should not exceed 50 μmol/l otherwise the risk of oxalosis developing is high.

Figure 4.6 shows the plasma oxalate levels of subjects with various disturbances of oxalate metabolism but with normal glomerular function, that is with plasma creatinine below 130 μmol/l. It can be seen that levels are clearly raised above normal in primary hyperoxaluria, slightly but definitely raised in mild metabolic hyperoxaluria and marginally raised in idiopathic hypercalciuria and secondary hyperoxaluria.

Circadian Variation of Plasma Oxalate. Implications for Oxalate Clearance

With the advent of a routine and reliable method (Kasidas and Rose 1986) for measuring oxalate levels in plasma in these laboratories, it became possible to

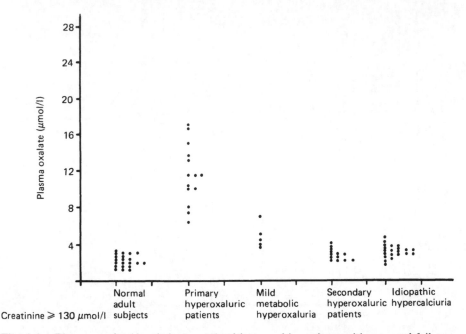

Fig. 4.6. Plasma oxalate levels in normal subjects and in patients without renal failure but with various disturbances of oxalate metabolism. Reproduced with permission from Kasidas GP and Rose GA (1987). Pathogenese und Klinik der Harnsteine XII. Steinkopff Verlag, Darmstadt, pp 143–147.

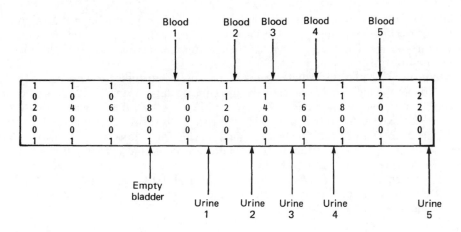

Fig. 4.7. Protocol for collections of blood and urine.

study circadian variation of plasma oxalate in 7 normal subjects using the protocol shown in Fig. 4.7.

Figure 4.8 shows the marked postprandial rises in oxalate in plasma in these subjects whilst they were on unrestricted dietary regimens. The plasma oxalate levels in these subjects progressively rose during the daytime hours varying

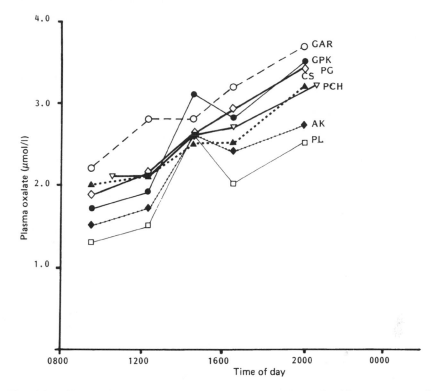

Fig. 4.8. Circadian variation of plasma oxalate in 7 normal subjects on unrestricted diets.

between 1.6 and 3.8 μmol/l and presumably they returned during the night to the lower levels found in the morning samples.

Similar diurnal variation has previously been seen in urine (Hargreave et al. 1977; Tiselius and Almgard 1977; Vahlensieck et al. 1982) with levels rising during waking hours and subsequently restored to basal levels during sleep (Hargreave et al. 1977). Dietary contribution of oxalate is highly variable and dependent upon the ingestion of certain foods such as tea, spinach, rhubarb, peanuts and chocolates (Finch et al. 1981) which are rich sources of oxalate (Kasidas and Rose 1980). Ingestion of oxalate-containing foods whilst subjects are on unrestricted diets leads to absorption and enhanced renal excretion of oxalate in normal subjects (Finch et al. 1981) as shown here in Fig. 4.9. Thus it can be assumed that ingestion of foods containing oxalate (such as tea), during the day could be responsible for the progressive increase in plasma oxalate concentrations. The observed rise in levels of urinary oxalate during daytime hours was attributed to the increased dietary intake of this anion during the waking hours. In Fig. 4.10 it can be seen that both restriction of oxalate in the diet and fasting prevented the variations and progressive daytime increase in oxalate in plasma. The levels of plasma oxalate during fasting or whilst subjects were on diets with no oxalate were much lower than those in samples taken from subjects on unrestricted diets and these values never rose above 2.3 μmol/l.

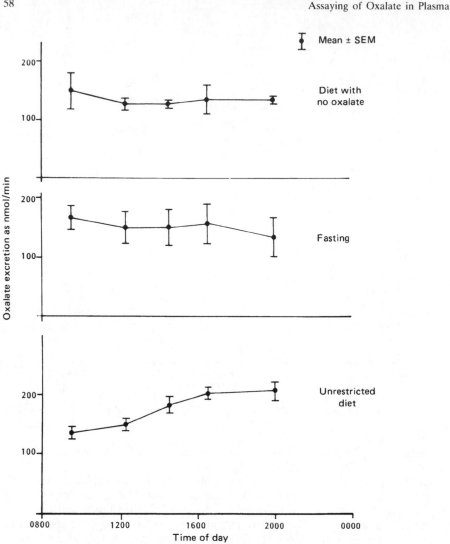

Fig. 4.9. Urine oxalate excretion rate in normal subjects on different dietary regimens. A steady gradual increase is seen during the day when subjects are kept on unrestricted diets.

In the light of the observed variations in the levels of plasma oxalate during the day, the values of oxalate clearances estimated by the same enzymatic technique in an earlier study (Kasidas and Rose 1986) may be doubtful since invariably these had been calculated from 24-h urine oxalate output and a single blood sample taken during mid-morning, a time when plasma oxalate is relatively low.

Timed urine collections were made during the above study. The blood samples obtained above were taken at mid-point intervals (see Fig. 4.7) of the timed urine collections. Measurement of urine volume and concentrations of

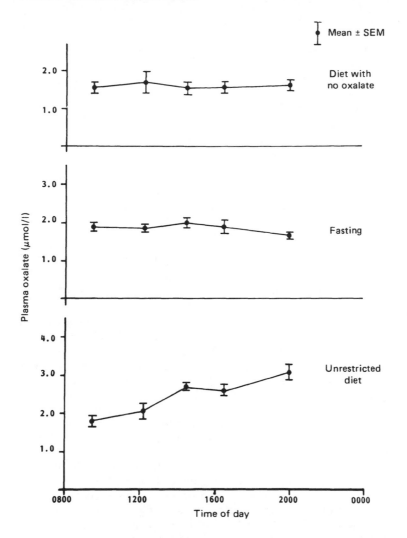

Fig. 4.10. Circadian variation of oxalate in plasma on different dietary schedules.

oxalate and creatinine in these samples enabled oxalate and creatinine clearances to be calculated. It can be seen in Fig. 4.11 that the majority of the estimates of oxalate/creatinine clearance ratios in this study were found to be less than unity. A mean ± SD of 0.76 ± 0.36 was found in a total of 97 estimates. However, in 15 of the estimates a mean ± SD of 1.42 ± 0.33 was found. In one subject during a fasting schedule estimates of oxalate/creatinine clearance ratios of less than unity were found which in a few hours changed to values greater than unity. No associations between diet, urine flow-rate, oxalate and creatinine clearances were found.

Oxalate/creatinine clearance ratios of much less than unity would imply net tubular reabsorption of oxalate whereas any values greater than unity would

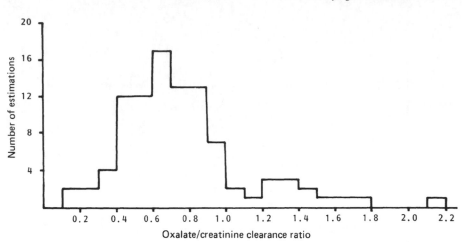

Fig. 4.11. Distribution of oxalate/creatinine clearance ratios in normal subjects.

imply net tubular secretion of oxalate because creatinine clearance itself exceeds the rate of glomerular filtration.

In man, Zarembski and Hodgkinson (1963) reported a net reabsorption of oxalate whereas others (Constable et al. 1979; Boer et al. 1984; Hodgkinson and Wilkinson 1974; Prenen et al. 1982; Williams et al. 1971) implied net secretion of oxalate from their findings of oxalate/creatinine clearance ratios greater than unity. For reasons discussed earlier in this chapter, the fluorimetric assay used by Zarembski and Hodgkinson (1963) for plasma oxalate gave erroneously high values of plasma oxalate and the oxalate clearance values calculated in that study seem doubtful. More recently Rehmert et al. (1983) using an enzymatic assay reported oxalate/creatinine clearance ratios below unity (range 0.09–0.33) in normal subjects.

Micropuncture studies in the rat (Greger et al. 1978) have indicated that oxalate is freely filterable at the glomerular site with a small but significant reabsorption of oxalate in the proximal nephron. They observed that at the proximal nephron and the pars recta oxalate is secreted to increase the tubular load to 124% of the filtered load. Oxalate clearance significantly ($p < 0.002$) exceeded that of inulin and even though the fractional clearance of oxalate was greater than unity over the whole range of diuresis they found a slight inverse correlation to urine flow rate. At high urinary flow rates they found that net secretion of oxalate decreased by 10% whereas antidiuresis produced net secretion of about 30% of the filtered load. Similar findings were reported, again in the rat, by Weinman et al. (1978).

Therefore, in the rat oxalate is handled by the kidney by complete glomerular filtration followed by bidirectional transport limited to the proximal tubule: there, a predominance of tubular secretion occurs to cause more than the filtered load to be excreted in the urine. Greger et al. (1978) suggested that reabsorption reflects a passive permeability of the nephrons to the small oxalate anion.

Osswald and Hautmann (1979) studied the renal handling of oxalate in 6 patients by a rapid injection intrarenal technique. They injected ^{14}C-labelled oxalate in the renal artery and studied the recovery characteristics of the tracer in urine. They used inulin as a glomerular marker and observed that the rate of excretion of oxalate tracer was 2.3 times greater than that of inulin. However, they observed a lower oxalate/inulin concentration ratio of 1.2 when 7 volunteers were given ^{14}C-oxalate intravenously. They attributed the differences to the larger amount of ionised oxalate entering the urine at a faster rate in the rapid intrarenal injection technique. They concluded that, in man, oxalate undergoes a bidirectional transport in the proximal tubule with an overall net secretion. They suggested that the reabsorption flux operates by a passive-diffusion mechanism whilst secretion is by an active saturable organic anion transport mechanism.

The mechanism of control of the processes of tubular reabsorption and secretion are not yet fully understood in the intact human kidney. It is not known whether the tubular cells themselves can produce oxalate metabolically or whether the presence of anions of other organic acids such as citrate, glycollate and urate in the tubular fluid can affect the transport of oxalate. Knight et al. (1979) showed that infusions of certain levels of para-amino-hippuric acid (PAH) can inhibit the transport of oxalate throughout the proximal tubule but had no effect on passive reabsorption of oxalate.

The inexplicable variations in the oxalate/creatinine clearance ratio of below and above unity observed here in the intact human kidney support the idea of bidirectional flux within the renal tubules in which either net tubular reabsorption or secretion can dominate.

Errors in the estimations of clearances arising from incomplete emptying of the bladder seem unlikely because this would equally have affected the creatinine excretion levels. Furthermore, subjects in this study had been advised to maintain a high intake of fluid to minimise the possible errors of bladder emptying. Mean ± SEM of flow rates of urine in the 97 estimations were 2.52 ± 0.22 ml/min. No definite association between urine flow rates and oxalate clearances were seen here in the intact human kidney. Further studies are warranted to elucidate the control mechanism and the factors which affect the tubular transport of oxalate.

Conclusions and Summary

Reliable and routine measurements of levels of oxalate in plasma are now possible. The discrepancy between in vivo radioisotopic and in vitro chemical/physical/enzymatic determinations of oxalate in plasma has now been resolved and it is becoming accepted that true normal plasma oxalate levels are between 1 and 3 μmol/l. Oxalogenesis and non-specificity of earlier in vitro chemical/physical/enzymatic assays contributed to the apparent higher levels obtained by these assays. Oxalogenesis during the post-sample collection period can now be controlled.

Circadian variation of plasma oxalate is quite large and this was shown to be diet-dependent. The collection of blood samples for plasma oxalate measurements should take account of this. Large variations of oxalate/creatinine clearance ratios have been observed. It seems that either net tubular secretion or reabsorption of oxalate can dominate in the intact human kidney. Further investigations are warranted to elucidate the control mechanism and the factors affecting the tubular transport of oxalate.

References

Ahmad S, Hatch M (1985) Hyperoxalaemia in renal failure and the role of hemoperfusion and hemodialysis in primary oxalosis. Nephron 41: 235–240

Akcay T, Rose GA (1980) The real and apparent plasma oxalate. Clin Chim Acta 101: 305–311

Barber HH, Gallimore EJ (1940) The metabolism of oxalic acid in the animal body. Biochem J 34: 144–148

Barrett JFB (1943) The oxalate content of blood. Biochem J 37: 254–256

Bennett DJ, Cole FE, Frohlich ED, Erwin DT (1979) A radioenzymatic isotope-dilution assay for oxalate in serum or plasma. Clin Chem 25: 1810–1813

Boer P, Van Leersum L, Endeman HJ (1984) Determination of plasma oxalate with oxalate oxidase. Clin Chim Acta 137: 53–60

Borland WW, Payton CD, Simpson K, MacDougall AI (1987) Serum oxalate in chronic renal failure. Nephron 45: 119–121

Carr PW, Bowers LD (1980) Principles of kinetic and equilibrium methods of analysis. In: Elving PJ, Winefordner JD, Kolthoff IM (eds) Immobilised enzymes in analytical and clinical chemistry. John Wiley & Sons, New York, pp 95–147

Chambers MM, Russell JC (1973) A specific assay for plasma oxalate. Clin Biochem 6: 22–28

Cole FE, Gladden KM, Bennett DJ, Erwin DT (1984) Human plasma oxalate concentration re-examined. Clin Chim Acta 139: 137–143

Constable AR, Joekes AM, Kasidas GP, O'Reagan P, Rose GA (1979) Plasma level and renal clearance of oxalate in normal subjects and in patients with primary hyperoxaluria or chronic renal failure or both. Clin Sci 56: 229–304

Crawhall JC, Watts RWE (1961) The oxalate content of human plasma. Clin Sci 20: 357–366

Endo R (1969) Rise and fall of oxalic acid in blood urine II Normal values. Med J Mutual Aid Assoc (Tokyo) 18: 254–256

Finch AM, Kasidas GP, Rose GA (1981) Urine composition in normal subjects after oral ingestion of oxalate rich foods. Clin Sci 60: 411–418

Flaschenträger B, Muller PB (1938) Zur biologie der oxalsäure 1. Microbestimmung der oxalsäure im Harn und Blut. Z Physiol Chem 251: 52–77

France NC, Windleborn EA, Wallace MR (1985) In vitro oxalogenesis and measurement of oxalate in serum. Clin Chem 31: 335–336

Gelot MA, Lavoue G, Belleville F, Nabet P (1980) Determination of oxalate in plasma and urine gas chromatography. Clin Chim Acta 106: 279–285

Greger R, Lang F, Oberleithner H, Deetjen P (1978) Handling of oxalate by the rat kidney. Pflugers Arch 374: 243–248

Hallson PC, Rose GA (1974) A simplified and rapid enzymatic method for determination of urinary oxalate. Clin Chim Acta 55: 29–39

Hargreave TB, Sali A, MacKay C, Sullivan M (1977) Diurnal variation in urinary oxalate. Br J Urol 49: 597–600

Hatch M, Bourke E, Costello J (1977) New enzymatic method for serum oxalate determination. Clin Chem 23: 76–78

Hodgkinson A, Wilkinson R (1974) Plasma oxalate concentration and renal excretion of oxalate in man. Clin Sci Molec Med 46: 61–73

Ichiyama A, Nakai E, Funai T, Oda T, Katafuchi R (1985) Spectrophotometric determination of oxalate in urine and plasma with oxalate oxidase. J Biochem (Tokyo) 98: 1375–1385

Jerez E (1986) Determination de acido oxalico en plasma por chromatografia liquid de atla efficacia. Rev Esp Fisiol 42: 441–448

Kasidas GP, Rose GA (1980) Oxalate content of some common foods; determination by an enzymatic method. J Human Nutr 35: 255–266

Kasidas GP, Rose GA (1986) Measurement of plasma oxalate in healthy subjects and in patients with chronic renal failure using immobilised oxalate oxidase. Clin Chim Acta 154: 49–58

Kasidas GP, Rose GA (1987) Measurement of plasma oxalate and when this is useful. In: Gasser G, Vahlensieck W (eds) Pathogenese und Klinik der Harnsteine XII. Steinkopff, Darmstadt, pp 143–147

Knight TF, Senekjian HO, Weinman EJ (1979) Effect of para-aminohippurate on renal transport of oxalate. Kidney Int 15: 38–42

Knowles CF, Hodgkinson A (1972) Automated enzymatic determination of oxalic acid in human serum. Analyst 97: 474–481

Kohlbecker G (1981) Direct spectrophotometric determination of serum and urinary oxalate with oxalate oxidase. J Clin Chem Clin Biochem 19: 1103–1106

Krugers Dagneaux PGLC, Elhorst JTK, Olthuis FMFG (1976) Oxalic acid determination in plasma. Clin Chim Acta 71: 319–325

Lopez M, Tuchman M, Scheinman JI (1985) Capillary gas chromatography measurement of oxalate in plasma and urine. Kidney Int 28: 82–84

Macaluso MP, Berg NO (1959) Calcium oxalate crystals in the kidney in acute tubular necrosis and other renal diseases with functional failure. Arch Path Microbiol Scand 46: 197–205

Merz KW, Maugeri S (1931) Über das Vorkommen and die Bestimmung der Oxalsäure im Blut. Physiol Chem 201: 31–37

Nuret P, Offner M (1978) A new method for determination of oxalate in blood serum by gas chromatography. Clin Chim Acta 82: 9–12

Osswald H, Hautmann R (1979) Renal elimination kinetics and plasma half life of oxalate in man. Urol Int 34: 440–450

Parkinson IS, Kealey T, Laker MF (1985) The determination of plasma oxalate concentrations using an enzyme/bioluminescent assay. Clin Chim Acta 152: 335–345

Pernet JL, Pernet A (1965) Dosage colorimetrique de l'acide oxalique dans les millieux biologiques. Ann Biol Clin 23: 1189–1207

Pinto B, Crespi G, Sole-Bacells F, Barcelo P (1974) Patterns of oxalate metabolism in recurrent stone formers. Kidney Int 5: 285–291

Prenen JAC, Boer P, Dorhout Mees EJ, Endeman HJ, Spoor SM, Oei HY (1982) Renal clearance of (^{14}C) oxalate: comparison of constant infusion with single injection techniques. Clin Sci 63: 47–51

O'Reagan PFB, Constable AR, Harrison AR, Joekes AM, Kasidas GP, Rose GA (1979) The management of the primary hyperoxaluric patient. In: Rose GA, Robertson WG, Watts RWE (eds) Oxalate in human biochemistry and clinical pathology. Wellcome Foundation, London, pp 209–219

Ramsay AG, Reed RG (1984) Oxalate removal by hemodialysis in end stage renal disease. Am J Kidney Dis 4: 123–127

Rehmert U, Wicher K, Ruge W, Bahlmann J (1983) Oxalate determination in plasma with oxalate oxidase. Investigation in healthy persons and patients with terminal renal failure before and after dialysis. Lab Med 7: 29–32

Rose GA (1985) Advances in analysis of urinary oxalate; the ascorbate problem solved. In: Schwille PO, Smith LH, Robertson WG, Vahlensieck W (eds) Urolithiasis and related clinical research. Plenum Press, New York, pp 637–644

Salyer WR, Keren D (1973) Oxalosis as a complication of chronic renal failure. Kidney Int 4: 61–66

Samsoondar J, Moore RW, Kellen JA (1983) Enzymatic determination of oxalates. Enzymes 30: 273–276

Sugiura M, Yamamura H, Hirano K et al. (1980) Enzymatic determination of serum oxalate. Clin Chim Acta 105: 393–399

Suzuki S (1934) The oxalic acid in blood. Jpn J Med Sci Biochem 2: 291–303

Thomsen A (1935) Über den oxalsäuregehalt des blutes. Hoppe-Seyler's Z Physiol Chem 237: 199–213

Tiselius HG, Almgard LE (1977) The diurnal urinary excretion of oxalate and the effect of pyridoxine and ascorbate on oxalate excretion. Eur Urol 3: 41–46

Tocco DJ, Duncan AFW, Noll RM, Duggan DE (1979) An electron capture procedure for the estimation of oxalic acid in urine. Anal Biochem 94: 470–476

Vahlensieck EW, Bach D, Hesse A (1982) Circadian rhythm of lithogenic substances in urine. Urol Res 10: 195–203

Weinman EJ, Frankfurt SJ, Ince A, Sansom S (1978) Renal tubular transport of organic acids. J Clin Invest 61: 801–806

Williams HE, Johnson GA, Smith LH (1971) The renal clearance of oxalate in normal subjects and in patients with primary hyperoxaluria. Clin Sci 41: 213–218

Wolthers BG, Hayer M (1982) The determination of oxalic acid in plasma and urine by means of capillary gas chromatography. Clin Chim Acta 120: 87–102

Zarembski PM, Hodgkinson A (1963) The renal clearance of oxalic acid in normal subjects and in patients with primary hyperoxlauria. Invest Urol 1: 87–93

Zarembski PM, Hodgkinson A (1965) The fluorimetric determination of oxalic acid in blood and other biological materials. Biochem J 96: 717–721

Zarembski PM, Hodgkinson A, Parsons FM (1966) Elevation of the concentration of plasma oxalic acid in renal failure. Nature 212: 511–512

Chapter 5

Primary Hyperoxaluria

R. W. E. Watts and M. A. Mansell

Clinical Aspects

The primary hyperoxalurias comprise three inborn errors of metabolism in which the urinary oxalate excretion is abnormally high (Table 5.1). Increases in the urinary oxalate excretion due to dietary and seasonal fluctuations in the intake and absorption of oxalate rarely cause the excretion of more than 0.5 mmol/24 h in adults. The three types of primary hyperoxaluria have the same

Table 5.1. The hyperoxalurias

Primary		
	Type I	Hyperoxaluria with hyperglycollic aciduria (Alanine:glyoxylate aminotransferase deficiency)
	Type II	Hyperoxaluria with L-glyceric aciduria (D-glycerate dehydrogenase deficiency)
	Type III	Intestinal (absorptive hyperoxaluria)
Secondary		
	Enteric:	Jejunoileal ileal bypass
		Small intestine resection
		Blind loops
		Diffuse disease of the small intestine (eg. Crohn's disease)
		Chronic pancreatic and biliary tract disease
Oxalate ingestion (acute poisoning)		
Excessive intake of ascorbic acid		
Ethylene glycol poisoning		
Adverse reaction to methoxyfluorane		
Glycine irrigation (after transurethral prostatectomy)		
Aspergillus infection		
Pyridoxine (Vitamin B_6) deficiency		

clinical phenotype. Type 1 occurs most commonly, and there are three clinically identifiable variants: infantile, juvenile and adult, recognisable by the age at which urolithiasis and/or renal failure begin. The juvenile group, with clinical presentation between 2 and about 18 years, is the most frequent subgroup. There have been too few reports of type II and type III patients to permit any similar classification for them by age of onset. Besides the individual patient's rate of oxalate production, the propensity to form stones will depend on urine concentration and the levels of the protective substances such as the pyrophosphate anion and glycosaminoglycans which normally inhibit crystal growth and aggregation in the urinary tract (Robertson et al. 1976). The glycosaminoglycans appear to be particularly important from this point of view with respect to calcium oxalate.

Some type I patients show either partial or complete correction of their excessive oxalate synthesis by pharmacological doses of pyridoxine (Watts et al. 1985a) indicating further genetic variation in this disorder. Patients with primary hyperoxaluria develop calcium oxalate crystals in many tissues (oxalosis) during the later stages of the disease, when oxalate retention due to renal failure is added to excessive oxalate de novo synthesis (see below). The oxalate deposits (Williams and Smith 1983) occur particularly in the tunica media of the muscular arteries and arterioles, myocardium, rete testis, bone marrow, calcifying regions of the bone parenchyma, synovia, retina and peripheral nerves (in this case the initial lesions are probably in the vasa nervora). In the absence of haemo- or peritoneal dialysis these oxalotic lesions usually remain clinically silent and they were first recognised in post-mortem material from patients dying in end-stage renal disease due to primary hyperoxaluria (Scowen et al. 1959). Standard haemo- and peritoneal dialysis regimes which control the uraemia remove oxalate relatively inefficiently (Watts et al. 1984), but do prolong life and give time for the clinical symptoms of oxalosis to develop. Dialysis regimes which would be unsuitable for elective longterm treatment are needed in order to keep up with the rate of oxalate synthesis in most cases of primary hyperoxaluria (Watts et al. 1984). The clinical and other features of oxalosis include: heart block, other cardiac dysrhythmias and myocarditis, ischaemic lesions on the extremities due to vascular insufficiency, osteosclerosis, synovitis, subcutaneous calcium oxalate calcinosis which may ulcerate through the skin, mononeuritis multiplex due to the involvement of the vasa nervora, and retinal lesions. The radiological findings, cardiomyopathy, arthritis and ocular manifestations of oxalosis were recently reviewed by Day et al. (1986), O'Callaghan et al. (1984), Hoffman et al. (1982) and by Meredith et al. (1984) respectively.

Biochemistry

General

Patients with one of the primary hyperoxalurias almost always excrete more than about 1 mmol oxalate/24 h and typically have raised plasma-oxalate concentrations even when the usual parameters of renal function are still within the normal range (Watts et al. 1983, 1985a). An example of the analytical data obtained for a clinically-stable 22-year-old man with primary hyperoxaluria type I is: GFR 51 ml/min/1.73 m², plasma oxalate 35 μmol/l, urine oxalate 4.25 mmol/24 h, urine glycollate 1.40 mmol/24h. This patient had deteriorated to the point of needing renal transplantation only 18 months after these observations were made. However, less severe cases are being increasingly recognised and some cases of idiopathic calcium-oxalate urolithiasis with urinary oxalate-excretion values which fluctuate around the upper limit of normal may represent mild genetic variants of one of the primary hyperoxalurias.

The proportion of the urinary oxalate, and by inference the plasma oxalate, which is derived from dietary sources will depend on the composition of the diet from both the viewpoint of its oxalate content and the amounts of substances which make oxalate insoluble in the gut. A value of about 10% has been widely quoted as the proportion of the urinary oxalate which is of dietary origins (Richardson and Farinelli 1981; Hesse and Bach 1982; Backman et al. 1985; Richardson 1986, personal communication). However, Rose (1982) found that the urinary excretion of oxalate by normal subjects fell to about 0.1 mmol/24 h when they took a low-oxalate diet. This suggests that more of the urinary oxalate is of dietary origin than has been previously suggested. In the case of primary hyperoxaluria types I and II, where the rate of oxalate synthesis is greatly augmented, dietary restriction has no appreciable effect on the urinary oxalate excretion. The main metabolic precursors of oxalate are glycine and ascorbate.

The excessive urinary oxalate excretion in types I and II primary hyperoxaluria is mainly of endogenous origin via glyoxylate, and although an excessive dietary intake of oxalate and ascorbate augment the urinary oxalate excretion, and one does advise their reduction in these disorders, the effect is usually small relative to the greatly augmented rate of oxalate synthesis via glyoxylate. This is in contradistinction to the situations where there is increased oxalate absorption and the restriction of oxalate intake and absorption, including the use of oxalate-binding agents (anion exchangers and calcium ions), is the central facet of therapy.

Glycollate links oxalate production to the generation of active glycolaldehyde which is transferred in the transketolase reaction and this in turn links it to the oxidative pathway of carbohydrate metabolism.

The Metabolic Lesion in Type I Primary Hyperoxaluria

Hydroxyproline, serine, the side-chains of the aromatic amino acids and glycollate are all minor metabolic precursors of oxalate (Richardson and Farinelli 1981; Richardson, personal communication). These each contribute 5% or less of the daily oxalate production. The metabolic pathway from serine to oxalate is via hydroxypyruvate rather than via ethanolamine (Richardson, personal communication). Negligible amounts of oxalate appear to be derived from carbohydrates and polyols under normal dietary conditions. The claim that the artificial sweetening agent diethylene glycol ($CH_2OH–CH_2–O–CH_2–CH_2OH$) is converted to oxalate in vivo has not been confirmed (K. E. Richardson, personal communication).

The conversion of glycine to glyoxylate is a peroxisomal oxidative deamination, generating hydrogen peroxide and catalysed by D-amino acid oxidase (glycine oxidase; EC 1.4.3.3.) (Ratner et al. 1944). The equilibrium of the peroxisomal pyridoxal 5'-phosphate-dependent alanine:glyoxylate aminotransferase (AGT; EC 2.6.1.44) is in the direction of glycine formation. Glyoxylate is also a substrate for the cytosolic glutamate:glyoxylate aminotransferase (GGT; EC 2.6.1.4) which is also pyridoxal 5'-phosphate-dependent (Rowsell et al. 1972; Danpure and Jennings 1986a). The mitochondrial metabolism of glyoxylate is a synergistic decarboxylation with 2-oxoglutarate (Crawhall and Watts 1962). Koch and Stokstad (1966) ascribed this reaction to the catalytic action of 2-oxoglutarate:glyoxylate carboligase. More recent work shows that this catalysis is a property of the first decarboxylating component of the 2-oxoglutarate dehydrogenase (EC 1.2.4.2) complex (Schlossberg et al. 1970; Hirashima et al. 1967). Oxalate appears to be formed from ascorbate non-enzymatically (see Chap. 2) and by a mechanism which does not involve glyoxylate. Although megadoses of ascorbate are well-known to increase oxalate production it has recently been shown that a conventional prophylactic dose of 100 mg daily does not produce a detectable effect on the plasma-oxalate concentration and other parameters of oxalate production in patients with end-stage renal disease and oxalate retention unrelated to primary hyperoxaluria (Morgan et al. 1988).

Koch et al. (1967) proposed that there were cytosolic and mitochondrial isoenzymes of 2-oxoglutarate: glyoxylate carboligase and that deficiency of the cytosolic isoenzyme was the metabolic lesion in primary hyperoxaluria type I. Recent studies using surgically excised human liver and modern differential centrifugation techniques have shown that 2-oxoglutarate:glyoxylate carboligase is wholly mitochondrial and that the apparent cytosolic catalytic activity was due to mitochondrial damage (Danpure et al. 1986). Primary hyperoxaluria type I is due to deficiency of peroxisomal alanine:glyoxylate aminotransferase (Danpure and Jennings 1986a; Danpure et al. 1987). Thus, primary hyperoxaluria is one of the expanding list of peroxisomal diseases (Table 5.2). The peroxisomes in liver biopsies from 4 patients with type I primary hyperoxaluria were reported to be of normal appearance although somewhat fewer and smaller than those in control liver biopsies (Iancu and Danpure 1987). The relationship between these differences and the metabolic lesion is unclear at present. Wanders et al. (1987) examined liver tissue from patients with Zellweger syndrome in which there are no identifiable peroxisomes and found

Table 5.2. Peroxisomal diseases

Zellweger syndrome
Pseudo-Zellweger syndrome
Adrenoleukodystrophy
Pseudo-neonatal adrenoleukodystrophy
Acatalasia
Infantile Refsum's disease
Refsum's disease (classical form)
Hyperpipecolic acidaemia
X-linked adrenoleukodystrophy
Chondrodysplasia punctatum rhizomelia
Primary hyperoxaluria type I

that the levels of alanine:glyoxylate aminotransferase were normal as were the patient's urinary excretions of oxalate and glycollate.

Defective transamination of glyoxylate to glycine by unfractionated homogenates of kidney tissues from two siblings with primary hyperoxaluria was reported 20 years ago (Dean et al. 1966, 1967) but discounted. Alanine:glyoxylate aminotransferase, which is also a serine:pyruvate aminotransferase (Danpure and Jennings 1986b), can be assayed on percutaneous needle biopsies of the liver. This provides an enzymological diagnosis of primary hyperoxaluria type I (Danpure et al. 1986, 1987). Evidence is accumulating that the degree of any residual catalytic activity measured on this unfractionated tissue approximately parallels the severity of the clinical phenotype (C. J. Danpure, personal communication).

The residual alanine:glyoxylate aminotransferase activity measured on the unfractionated liver biopsies is ascribed either to cross-specificity between glutamate:glyoxylate and alanine:glyoxylate aminotransferases with respect to glyoxylate and/or to residual activity of alanine:glyoxylate aminotransferase itself. The existence of pyridoxine-sensitive and pyridoxine-resistant variants of primary hyperoxaluria type I (Gibbs and Watts 1970; Watts et al. 1985a) can be explained by genetic heterogeneity with respect to mutations which change the affinity of the enzyme binding-site for the cofactor. Alternatively, the region of the molecule containing the cofactor-binding site may be deleted.

Nakatani et al. (1985) described the isolation, purification and enzyme-linked immunoadsorbent assay (ELISA) of human hepatic alanine:glyoxylate aminotransferase. Heterogeneity with respect to the presence or absence of immunologically detectable alanine:glyoxylate aminotransferase in different patients with primary hyperoxaluria type I has recently been demonstrated (Wise et al. 1987). The next step in the study of the biochemistry of primary hyperoxaluria type I will be to determine the amino-acid sequence of the human enzyme and its structure and to clone the gene directing its synthesis.

The Metabolic Lesion in Type II Primary Hyperoxaluria

Primary hyperoxaluria type II is due to deficiency of D-glycerate dehydrogenase (EC 1.1.1.29), which normally catalyses the reduction of hydroxypyruvate to D-glycerate. This metabolic lesion is compensated by the lactate-dehydrogenase (EC 1.1.1.27)-catalysed reduction of hydroxypyruvate to L-glycerate and this leads to L-glyceric aciduria. Williams and Smith (1971) proposed that this reduction is coupled to the lactate-dehydrogenase-catalysed oxidation of glyoxylate to oxalate through the $NAD^+/NADH$ shuttle. Thus, increased reduction of hydroxypyruvate shifts the equilibrium $NAD^+ \Leftrightarrow NADH$ in the direction of NAD^+ which in turn promotes the oxidation of glycollate to oxalate.

Primary Hyperoxaluria Type III

Patients with primary hyperoxaluria type III do not have any associated hyperglycollic aciduria or L-glyceric aciduria. Their urinary oxalate excretion (usually about 1–2 mmol/day) is similar to that of some patients with the other types of primary hyperoxaluria. The underlying abnormality appears to be a primary hyperabsorption of oxalate in the absence of either overt or covert intestinal disease. Oxalate absorption has been regarded as a passive non-carrier-mediated process. However, Hatch et al. (1984) presented evidence for carrier-mediated active transport of oxalate in the rabbit colon and Knichelbein et al. (1986) showed oxalate–chloride exchange across rabbit ileal brush border membranes. The metabolic lesion in primary hyperoxaluria type III *might* involve one of these processes. Yendt and Cohanim (1986), whose patient excreted about 1 mmol of oxalate/24 h, reported mildly increased absorption and urinary excretion of calcium and magnesium. Increases in the rate of urinary oxalate excretion due to dietary and seasonal fluctuations in the intake and absorption of calcium and oxalate, or in some cases of idiopathic hypercalciuria, rarely cause oxalate excretion values greater than 0.5 mmol/24h. Yendt and Cohanim (1986) showed that the urinary excretion of oxalate was reduced by a low oxalate diet and by hydrochlorothiazide.

Investigation and Evaluation of Patients with Primary Hyperoxaluria

Patients with primary hyperoxaluria usually present with stones and/or renal failure. Some are detected when they are still asymptomatic and a few are not identified until they have manifestations due to oxalosis. Their complete evaluation requires: (i) measurement of blood and urine oxalate, glycollate and L-glycerate levels, (ii) enzyme assays, (iii) the identification and evaluation of oxalotic organ damage. Measurements of the size of the oxalate metabolic pool, of the accretion rates of tissue oxalate and of the total production rate of oxalate also give additional information.

Primary hyperoxaluria types I and II are diagnosed initially on the basis of the pattern of urinary organic acid excretion. Specific enzymatic methods are now available for the measurement of urinary and plasma oxalate (see Chaps 2 and 4). These have been automated and are suitable for making sequential measurements in order to assess progress (Kasidas and Rose 1985, 1986). Glycollate can be measured enzymatically (Kasidas and Rose 1979), and gas chromatographic methods with capillary columns give good resolution and permit the simultaneous assay of D-glycerate, oxalate and glycollate (Chalmers et al. 1984). HPLC methods are also available.

The diagnosis of type III primary hyperoxaluria depends on the demonstration of hyperoxaluria without either glycollic aciduria or L-glyceric aciduria, the absence of clinical and biochemical evidence of generalised intestinal malabsorption and excessive translocation of a standard dose of ^{14}C-labelled oxalate from the gut into the urine (Chadwick et al. 1973; Williams 1976). There is evidence that the hyperabsorption of oxalate in primary hyperoxaluria type III is associated with increased absorption of calcium and magnesium (Yendt and Cohanim 1986). Hyperabsorption of oxalate has also been invoked to explain the much smaller degrees of hyperoxaluria which are observed in some cases of idiopathic hypercalciuria (Marangella et al. 1982).

The definitive enzymological diagnoses of types I and II primary hyperoxaluria are made on percutaneous liver biopsy tissue (Danpure et al. 1987) and peripheral blood leukocytes respectively (Williams and Smith 1971). The radiochemical assay for alanine:glyoxylate aminotransferase has recently been miniaturised so that it requires only 100 μg tissue and could therefore be used on a fetal liver biopsy for prenatal diagnosis (Allsop et al. 1987).

Patients with type I primary hyperoxaluria have been studied in the greatest detail from the pathophysiological viewpoint, in order to evaluate the relative roles of oxalate overproduction and oxalate retention in the production of oxalosis and the effects of different treatment modalities. The size of the metabolic pool of oxalate, the accumulation rate of tissue oxalate and the total production rate of oxalate are derived from measurements of plasma and urine carbon-14 and technetium-99m after a single bolus injection of ^{14}C-labelled oxalate and ^{99m}Tc-labelled diethylenetriaminepentaacetate (DTPA) with timed sequential blood sampling and continuous urine collection by voluntary voiding (Watts et al. 1983, 1984, 1985a). The total plasma clearance and equilibrium distribution volumes for both tracers are calculated (Veall and Gibbs 1982). An input pulse followed by a series of regular unit pulses is regarded as being mathematically equivalent to a priming dose followed by a continuous infusion. The observed plasma time-curve is convolved with this input function, the size of the initial pulse is determined by iteration to minimise the slope of the terminal part of the notional plasma radioactivity function. This provides a computer simulation of the data which would have been obtained from a continuous infusion study with an almost perfectly optimised priming dose. The clearance and distribution volumes are then derived in the conventional way (Constable et al. 1979). The experimentally determined and derived data are listed in Table 5.3.

Table 5.3. Measured and derived values used to assess the propensity to systemic oxalosis in patients with primary hyperoxaluria

Measured values
[99m]Tc-labelled diethylenetriaminepenta-acetate (DTPA) renal clearance (ml/min) [GFR]
[99m]Tc-labelled DTPA distribution volume (l) (extracellular fluid space) [ECF]
Plasma ^{14}C-labelled oxalate clearance (ml/min) [PC_{Ox}]
[14]C-labelled oxalate distribution volume (l) [OxDV]
Renal ^{14}C-labelled oxalate clearance (ml/min) [RC_{Ox}]
Urinary oxalate concentration (by enzymatic assay) (μmol/l) [U_{Ox}]

Derived values

Plasma oxalate (μmol/l)[a] [P_{Ox}]	= ([U_{Ox}] × urine flow rate)/[RC_{Ox}]
Oxalate metabolic pool size (μmol) [OxMP]	= [OxDV] × [P_{Ox}]
Plasma total oxalate clearance (μmol/day) (= oxalate production rate) [$P\Sigma C_{Ox}$]	= [PC_{Ox}] × [P_{Ox}] × 1.44
Tissue oxalate clearance (ml/min) [TC_{Ox}]	= [PC_{Ox}] − [RC_{Ox}]
Tissue oxalate accumulation (μmol/day) [TOxA]	= [TC_{Ox}] × [P_{Ox}] × 1.44

[a]The reference range obtained by enzymatic assay (Kasidas and Rose 1986) is 1.3–3.1 μmol/l and close to the reference range for the derived value (0.65–1.45 μmol/l).

The whole-body metabolism of oxalate can be regarded as a single compartment system which oxalate enters from dietary and biosynthetic sources and which it normally leaves by excretion into the urine. Impaired renal function causes oxalate retention with expansion of the size of the metabolic pool of oxalate and accretion of oxalate in the tissues. Although Watts et al. (1984) reported a very slight conversion of ^{14}C-labelled oxalate to respiratory $^{14}CO_2$, equivalent to a plasma-oxalate clearance of 0.1 ml/min in a patient with gross oxalate overproduction and retention, it was within the limits of the manufacturers stated 99% radiochemical purity of the ^{14}C-labelled oxalate used. A trace degree of radiochemical contamination was subsequently shown to have been responsible. This supports the generally held view that oxalate is not endogenously catabolised in man. There was also no detectable conversion of injected ^{14}C-labelled oxalate to a radioactive faecal metabolite (Watts et al. 1984). Elder and Wyngaarden (1960) and Hodgkinson and Wilkinson (1974) showed that virtually the whole of an intravenous injection of ^{14}C-labelled oxalate was excreted in the urine within two days. This supports the view that excretion of ^{14}C-labelled oxalate into the gut and subsequent degradation by the newly discovered organism *Oxalobacter formigenes* (Allison et al. 1986) in the human colon is unlikely to be a factor complicating the interpretation of data derived from the in vivo dynamic studies with ^{14}C-labelled oxalate. Loss of *Oxalobacter formigenes* from the faeces may be a factor in the excessive absorption of oxalate from the colon which occurs in patients with enteric hyperoxaluria (Allison et al. 1986).

Table 5.4 shows the results of dynamic studies with ^{14}C-labelled oxalate in 3 patients who were treated with pharmacological doses of pyridoxine. Patient 1 was fully responsive, Patient 2 was partly responsive and Patient 3 was completely resistant demonstrating this aspect of the genetic heterogeneity of the disease.

The evaluation of patients with primary hyperoxaluria will in future involve assessing the extent of oxalosis and the severity of its effect on organ function in greater detail than has been customary heretofore. This will require imaging techniques, biopsies and tests of function directed particularly at the cardiovascular, pulmonary, peripheral nervous and skeletal systems.

Table 5.4. The effect of pharmacological doses of pyridoxine in three patients with primary hyperoxaluria (Watts, Veall, Purkiss et al. 1985a)

	Patient 1		Patient 2		Patient 3		Reference ranges
	On B$_6$ (450 mg/day)	Off B$_6$	On B$_6$ (800 mg/day)	Off B$_6$	On B$_6$ (800 mg/day)	Off B$_6$	
Plasma oxalate concentration (μmol/l)	0.597	4.71	3.78	43.4	35.1	35.3	0.65–1.48
Oxalate metabolic pool size (μmol)	11.3	82.8	64.6	686	576	671	10.0 –33.2
Tissue oxalate accretion rate (μmol/24h)	3.0	0	28.8	1740	354	976	0 –24
Urinary oxalate excretion (μmol/24h)	150	1220	324	3280	4630	4680	156 –282

Table 5.5. The relevance of different treatment modalities to the different types of primary hyperoxaluria

	Type		
	I	II	III
Diet Low oxalate	+	+	+
Low calcium	+	+	−
Low vitamin C	+	+	+
Low vitamin D	+	+	+
Hydration	+	+	+
Inhibition of crystal growth (administration of MgO, Mg(OH)$_2$ or orthophosphate)	+	+	+
Oxalate binding agents	−	−	+
Thiazides	−	−	+
Pyridoxine	+	−	−
Dialysis	+	+	+
Renal transplantation	+	+	+
Hepatic transplantation	+	−	−

Treatment

The different treatment modalities that have to be considered in each type of primary hyperoxaluria are summarised in Table 5.5. All patients with urinary stone disease should drink sufficient fluid to maintain a measured urine volume of at least 3 1/24 h. Dietary measures (Tables 5.6 and 5.7) are recommended in all three types although their main value is in type III. The intakes of both calcium and oxalate are restricted in types I and II in order to reduce the dietary contributions of both of these ions to their concentrations in the urine and to correct the tendency for reduced calcium intake to increase oxalate absorption. Table 5.6 shows the dietary advice for patients with types I and II primary hyperoxaluria. The absorption of oxalate is reduced by increasing the ionised calcium intake (Williams 1976) and by anion exchangers such as cholestyramine. Yendt and Cohanim (1986) reported that hydrochlorothiazide lowered the urinary oxalate excretion in their patient with primary hyperoxaluria type III and they attributed this to the inhibitory action of the drug on calcium absorption. Magnesium ions and the increased level of urinary pyrophosphate produced by feeding orthophosphate increase the physiological crystallisation-inhibitory potency of the urine. Pharmacological doses of pyridoxine reduce the excretion of urinary oxalate in some patients with primary hyperoxaluria type I. This possibility should be explored as part of the initial evaluation of the patient.

Removal of oxalate by haemo- or peritoneal dialysis or by haemofiltration cannot keep up with the rate of oxalate synthesis in patients with primary hyperoxaluria and advanced renal failure (Watts et al. 1984). Therefore, the treatment of end-stage renal failure due to primary hyperoxaluria type I presents special problems because the metabolic pool of oxalate expands

Table 5.6. Diet for patients with primary hyperoxaluria. Low oxalate (approximately 0.77 mmol ($\equiv$70 mg) $H_2C_2O_4$) and low calcium (approximately 17.5 mmol (700 mg) Ca) with minimal intakes of vitamins C and D (personal communication from Miss P. Hulme, BSc, SRD)

Breakfast
Porridge or cornflakes with milk from allowance, sugar if desired
Bread, butter and marmalade
Egg, bacon or sausage or fish (if desired)
Weak coffee with milk from allowance, sugar if desired
or fruit squash low in vitamin C

Mid-morning
Weak coffee with milk from allowance, sugar if desired
or fruit squash low in vitamin C
Plain biscuits if desired

Mid-day or evening
Meat, poultry, fish or egg dish
Average helping of cooked vegetables or salad (see list to avoid)
Potatoes or rice or spaghetti
Fruit, fresh or tinned (see list to avoid)

Tea
Weak coffee with milk from allowance, sugar if liked
or fruit squash low in vitamin C *or* fizzy drink if wished

Evening or mid-day
Meat, Poultry, fish or egg dish
Vegetables as at mid-day
Bread or potatoes if liked
Fruit as mid-day

Bedtime
Weak coffee with milk from allowance, sugar if liked
or fruit squash low in vitamin C

Daily allowances
150 ml ($\frac{1}{4}$ pint) milk daily
4 slices white bread[a]
2 slices wholemeal bread
1 egg
6 cups weak coffee

---

[a]In place of 1 slice of white bread you make take either 1 oz pastry in a pie or tart or 4 small biscuits.

Notes
1. Drink at least 3 l (5 pints) of fluid daily. Water may be flavoured with a low-vitamin-C squash if you wish. You must drink enough to ensure that you pass 3 l (about 5 pints) of urine each 24 h. Always drink about 300 ml (about $\frac{1}{2}$ pint) of water before going to bed. Measure the total volume of urine that you pass in a day to check that you are drinking enough. You need extra drinks in hot weather
2. Do *not* take any vitamin-C or D containing medicines

Table 5.7. Foods which are either allowed or not allowed in primary hyperoxaluria

	Foods allowed	Foods not allowed
Meat and fish	Meat of any kind White fish (no bones)	Fish with fine bones eg. sardines, pilchards, kippers, herrings, tinned salmon, shell-fish, whitebait, fish paste
Milk products	Milk (see daily allowance Butter)	Cream, condensed milk, evaporated milk, dried milk, yoghurt, ice-cream, cheese – all kinds
Bread and cakes	Bread (see daily allowance or exchange)	Wholemeal biscuits, eg. digestive, shop-bought cakes, biscuits and puddings
Cereals	Cornflakes, Rice Krispies, Weetabix, porridge made with water, arrowroot, cornflour, oatmeal, macaroni, rice, spaghetti	All wholemeal breakfast cereals, eg. All Bran, Grape Nuts etc. Muesli
Fruit	Fresh, stewed or tinned fruit except those listed to avoid	Gooseberries, strawberries, raspberries, blackcurrants, rhubarb, cranberries, lemons, tangerines, oranges, grapefruit, all dried fruit and nuts
Vegetables	Most vegetables, but see those to avoid	All beans, broccoli, spinach, beetroot, watercress, parsley
Beverages	Coffee (see daily allowance) instant or infused freshly ground, some herbal teas may be suitable (ask dietitian), fruit squashes, fizzy drinks, eg. Lucozade, lemonade	Tea, cocoa, Ovaltine, Bournvita, Horlicks, drinking chocolate, orange and grapefruit juices (fresh, canned or carton), fruit squashes with added vitamin C, Ribena Alcohol: ask your doctor about this
Miscellaneous	Coffeemate milk substitute, boiled sweets, clear toffee, sugar, glucose, jams, honey, jelly, pickles, Bovril, fats and oils	Fruit gums, plain and milk chocolate, carob black treacle, sauces and soups made with milk

Avoid all wrapped foods where milk powder, calcium salts, or the additives E282 (calcium propionate) and E516 (calcium sulphate) are listed as ingredients

acutely and, when oxalate retention is combined with oxalate overproduction, potentially lethal oxalosis progresses rapidly. Before this stage, overall renal function usually declines slowly with exacerbations associated with episodes of obstructive uropathy. The terminal uraemic illness is characteristically short. Patients with primary hyperoxaluria whose uraemia is controlled by conventional dialysis regimes may appear deceptively well until they develop clinical manifestations of generalised oxalosis.

The results of renal transplantation in pyridoxine-resistant primary hyperoxaluria type I, the variant which has been most studied, have been generally poor when it has been introduced according to the criteria used in other types of renal disease (Klauwers et al. 1968; Deodhar et al. 1969; Chesney et al. 1984; Vanrenterghem et al. 1984). However, some exceptions to this generalisation have been reported (David et al. 1983; Whelchel et al. 1983; Scheinman et al. 1984). Intratubular and interstitial deposition of calcium oxalate in the transplanted kidney is a hazard specific to the primary hyperoxalurias. Delayed onset of renal function, perioperative underhydration and the presence of a large filtered load derived from the expanded rapidly exchangeable pool of accumulated oxalate predispose to this complication.

Live related donors have been thought to offer the best prospect of success, but this is not necessarily so (Watts et al. 1988). Apart from the timing of the operation (Morgan et al. 1987; Watts et al. 1988) vigorous perioperative haemodialysis and the maintenance of high rates of urine flow by hydration are the most important factors favouring graft function in pyridoxine-resistant primary hyperoxaluria type I. The value of orthophosphate, magnesium ions, thiazides (Scheinman et al. 1984), etidronate (David et al. 1983) and pretransplantation parathyroidectomy (David et al. 1983) is more difficult to evaluate. The same considerations apply to the other types of primary hyperoxaluria once they have reached end-stage renal failure.

Recent work by Morgan et al. (1987) shows why the timing of renal transplantation in relation to the level of overall renal function should be an important determinant of the longterm success of renal transplantation in these patients. Oxalate retention occurs late in the evolution of renal failure due to causes other than primary hyperoxaluria but in this condition it becomes appreciable when the GFR has fallen only a little below the normal range. Figure 5.1 shows the effect of decreasing GFR on the parameters which measure the patients' predisposition to oxalosis. Until very recently, most primary hyperoxaluric patients were transplanted when they had a negligible GFR and were being treated by conventional haemo- or peritoneal dialysis regimes. Under these circumstances, the body burden of oxalate will be very great, a large amount of oxalate will enter the newly grafted kidney and this may precipitate in the renal tubules and interstitium producing permanent damage. Morgan and his colleagues (1987) conclude that renal transplantation should be performed when the GFR is still above the value at which a further fall is associated with rapid increases in plasma oxalate, the size of the metabolic pool of oxalate and the accretion rate of tissue oxalate (Fig. 5.1). The institution of auxiliary measures to remove oxalate from the body should be considered when the GFR falls below 40 ml/min/1.73 m^2, with a view to their introduction when the GFR is in the range 20–25 ml/min/1.73 m^2 and the run-up period to renal transplantation begins. Overall renal function starts to decline very rapidly in primary hyperoxaluria when the GFR falls below about 20 ml/min/1.73 m^2 and it is wise to begin *planning* for a transplant well in advance of its being needed. Haemodialysis should be vigorous, for example 6 h/day for 6 days/week, during the preoperative and perioperative periods and a very high rate of urine flow, up to 20 l per day if possible, should be maintained. Sequential measurements of plasma oxalate concentration should be used to monitor the

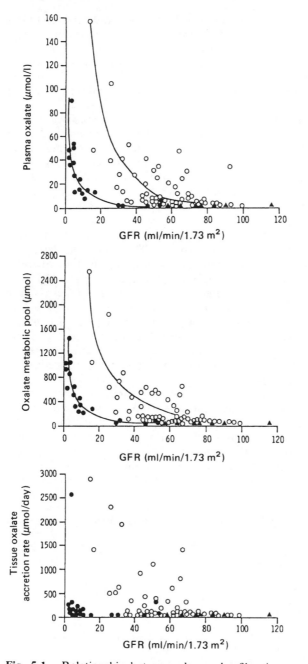

Fig. 5.1. Relationship between glomerular filtration rate (GFR) and plasma oxalate concentration (P_{Ox}), oxalate metabolic pool size (OxMP) and tissue oxalate accretion rate (TOxA). Filled circles, patients with renal failure unrelated to primary hyperoxaluria; open circles, patients with primary hyperoxaluria; filled triangles, healthy control subjects. Reproduced from Morgan et al. 1987, with permission of the authors and publisher (S. Karger AG).

biochemical progress. This can be supplemented by dynamic radiotracer studies to measure the size of the metabolic pool of oxalate and the accretion rate of tissue oxalate at less frequent intervals, although these are not essential for routine effective management. Pharmacological doses of pyridoxine should be given to pyridoxine-responsive patients even if they are only partially responsive and to those whose pyridoxine responsiveness has not been assessed preoperatively.

Watts et al. (1988) compared the outcome in two patients who were transplanted as soon as possible after they presented with *incipient* end-stage renal failure with the results in four others who were transplanted after prolonged periods of haemo- or peritoneal dialysis. The two early transplanted patients were fully active and working with GFRs of 57 and 41 ml/min/1.73 m^2, 19 and 17 months postoperatively respectively, whereas the other 4 patients' transplants failed between 6 and 11 months postoperatively due to oxalate deposition in the grafts.

Alanine:glyoxylate aminotransferase is virtually only expressed in the liver where, in man, it is confined to the peroxisomes. The metabolic lesion in primary hyperoxaluria type I has been corrected and renal function restored by combined hepatic and renal transplantation (Watts et al. 1985b, 1987). The liver transplant replaces the missing or defective enzyme by the corresponding normal enzyme which is in the physiological organ, cell and subcellular organelle as well as being correctly orientated with respect to other enzyme systems and to its substrate. Four patients with pyridoxine-resistant primary hyperoxaluria type I have now (October 1987) been treated by hepatorenal transplantation in the UK. The results are shown in Table 5.8. Two other patients are known to have also been successfully operated on, one we know positively to show full biochemical correction and data on the second are awaited. It is noteworthy that the urinary glycollate levels decrease into the normal or subnormal range immediately after the operation whereas the oxalate levels decline more slowly. This is attributed to the gradual mobilisation of oxalate from the greatly expanded metabolic pool of oxalate and from the generalised oxalotic deposits, which occurs when the excessive oxalate synthesis is abolished and the body fluids become progressively less saturated with respect to calcium oxalate. The

Table 5.8. Combined hepatic and renal transplantation for the treatment of pyridoxine-resistant primary hyperoxaluria type I

Patient	Outcome	Reference
NA	Biochemical correction: died 2 months postoperation (generalised cytomegalovirus infection)	Watts et al. 1985b
BG	Died in the immediate postoperative period (severe generalised oxalosis)	Calne, Rolles and Watts (unpublished data)
GF	Biochemical correction: full activity, working 11 months postoperatively	Watts et al. 1987
HK	Biochemical correction: full activity at home 5 months postoperatively	Watts, Baker, Calne and Rolles (unpublished data)

glycollate anion does not form an insoluble calcium salt so that it does not accumulate in the tissues. The initial strategy was to transplant the liver and kidney at the same operation using the two organs from the same donor. The most recent cases have been done in two stages, the liver transplantation first followed by the renal transplantation several weeks later, after a period of very vigorous haemodialysis to deplete the oxalate metabolic pool. This should reduce the oxalate load presented to the newly grafted kidney and improve its chances of survival. The relative merits of these two approaches remain to be assessed in detail.

The treatment of primary hyperoxaluria currently raises five issues: (1) general measures to minimise stone formation, (2) local management of stones by lithotripsy, percutaeneous and endoscopic surgery and by open operation, (3) the selection of patients for pyridoxine, (4) the timing of renal transplantation and the associated aspects of management which are necessary in order to minimse the risk of generalised oxalosis becoming the main prognostic factor, (5) the place of hepatorenal transplantation, whether this should be offered as definitive treatment once renal failure approaches or held in reserve until the patient has had one or more failed renal transplants.

References

Allison MG, Cook HM, Milne DB, Gallagher S, Clayman RV (1986) Oxalate degradation by gastrointestinal bacteria in humans. J Nutr 116: 455–460

Allsop J, Jennings PR, Danpure CJ (1987) A new microassay for human liver alanine:glyoxylate aminotransferase. Clin Chim Acta (in press)

Backman U, Danielson BG, Ljunghall S (1985) Renal stones. Aetiology, management, treatment. Almquist and Wicksell, Stockholm p. 78

Chadwick VS, Modha K, Dowling RH (1973) Mechanism for hyperoxaluria in patients with ileal dysfunction. N Engl J Med 289: 172–176

Chalmers RA, Tracey BM, Mistry J, Griffiths KD, Green A, Winterborn MH (1984) L-glyceric aciduria (primary hyperoxaluria type II) in siblings in two unrelated families. J Inher Metab Dis 7 (Suppl 2): 133–134

Chesney RW, Friedman AL, Breed AL, Adams ND, Lemann J (1984) Renal transplantation in primary oxaluria. J Paediatr 104: 322–323

Constable AR, Joekes AM, Kasidas GP, O'Regan P, Rose GA (1979) Plasma level and renal clearance of oxalate in normal subjects and in patients with primary hyperoxaluria or chronic renal failure or both. Clin Sci 56: 299–304

Crawhall JC, Watts RWE (1962) The metabolism of glyoxylate by human and rat liver mitochondria. Biochem J 85: 163–171

Danpure CJ, Purkiss P, Jennings PR, Watts RWE (1986) Mitochondrial damage and the subcellular distribution of 2-oxoglutarate:glyoxylate carboligase in normal human and rat liver and the liver of a patient with primary hyperoxaluria type I. Clin Sci 74: 417–425

Danpure CJ, Jennings PR (1986a) Peroxisomal alanine: glyoxylate aminotransferase deficiency in primary hyperoxaluria type I. FEBS Lett 201: 20–24

Danpure CJ, Jennings PR (1986b) Alanine:glyoxylate and serine:pyruvate aminotransferases in primary hyperoxaluria type I. Biochem Soc Trans 14: 1059–1060

Danpure CJ, Jennings PR, Watts RWE (1987) The enzymological diagnosis of primary hyperoxaluria type I by measuring the alanine:glyoxylate aminotransferase activity in hepatic percutaneous needle biopsies. Lancet I: 289–291

David DS, Cheigh JS, Stenbzel KH, Rubin AL (1983) Successful renal transplantation in a patient with primary hyperoxaluria. Transplant Proc 15: 2168–2171

Day DL, Scheinman JI, Mahan J (1986) Radiological aspects of primary hyperoxaluria. Am J Radiol 146: 395–401

Dean BM, Griffin WJ, Watts RWE (1966) Primary hyperoxaluria. The demonstration of a metabolic abnormality in kidney tissue. Lancet I: 406

Dean BM, Watts RWE, Westwick WJ (1967) Metabolism of [1-^{14}C]glyoxylate, [1-^{14}C]glycollate, [1-^{14}C]glycine and [2-^{14}C]glycine by homogenates of kidney and liver tissue from hyperoxaluric and control subjects. Biochem J 105: 701–707

Deodhar SD, Tung KSK, Zulke V, Nakamoto S (1969) Renal homotransplantation in a patient with primary familial oxalosis. Arch Pathol 87: 118–124

Elder TD, Wyngaarden JC (1960) The biosynthesis and turnover of oxalate in normal and hyperoxaluric subjects. J Clin Invest 39: 1337–1344

Gibbs DA, Watts RWE (1970) The action of pyridoxine in primary hyperoxaluria. Clin Sci 38: 277–286

Hatch MR, Freel RW, Goldner AM, Earnest DL (1984) Oxalate and chloride absorption by the rabbit colon: sensitive to metabolic and anion transport inhibitors. Gut 25: 232–237

Hesse von A, Bach D (1982) Harnsteine Pathobiochemie und klinisch-chemie Diagnostik. In: Breuer H, Büttner H, Stamm D (eds) Klinische Chemie in Einzeldarstellungen Band 5. Georg Thieme, Stuttgart, pp 33–40

Hirashima M, Hayakawa T, Koike H (1967) Mammalian α-keto acid dehydrogenase complexes. 2. An improved procedure for the preparation of 2-oxo-glutarate dehydrogenase complex from pig heart muscle. J Biol Chem 242: 902–907

Hodgkinson A, Wilkinson R (1974) Plasma oxalate concentration and renal excretion of oxalate in man. Clin Sci 46: 61–73

Hoffman GS, Schumacher HR, Paul H, Cherian V, Reed R, Ramsay AG, Franck WA (1982) Calcium oxalate microcrystalline-associated arthritis in end-stage renal disease. Ann Int Med 97: 36–42

Iancu TC, Danpure CJ (1987) Primary hyperoxaluria type I: ultrastructural observations in liver biopsies. J Inher Metab Dis 10: 330–338

Kasidas GP, Rose GA (1979) A new enzymatic method for the determination of glycollate in urine and plasma. Clin Chim Acta 96: 25–36

Kasidas GP, Rose GA (1985) Continuous-flow assay for urinary oxalate using immobilised oxalate oxidase. Ann Clin Biochem 22: 412–419

Kasidas GP, Rose GA (1986) Measurement of plasma oxalate in healthy subjects and in patients with chronic renal failure using immobilised oxalate oxidase. Clin Chim Acta 154: 49–58

Klauwers J, Wolff PL, Cohn R (1968) Failure of renal transplantation in primary oxalosis. JAMA 209: 551

Knichelbein RG, Aronson PS, Dobbins JW (1986) Oxalate transport by anion exchange across the rat ileal brush border. J Clin Invest 77: 170–175

Koch J, Stokstad ELR (1966) Partial purification of a 2-oxoglutarate:glyoxylate carboligase from rat liver mitochondria. Biochem Biophy Res Comm 23: 585–596

Koch J, Stokstad ELR, Williams HE, Smith LH (1967) Deficiency of 2-oxoglutarate:glyoxylate carboligase activity in primary hyperoxaluria. Proc Natl Acad Sci USA 57: 1123–1129

Marangella M, Fruttero B, Bruno M, Linari F (1982) Hyperoxaluria in idiopathic calcium stone disease: further evidence of intestinal hyperabsorption of oxalate. Clin Sci 63: 381–385

Meredith TA, Wright JD, Gammon JA, Fellner SK, Warshaw BL, Maio M (1984) Ocular involvement in primary hyperoxaluria. Arch Ophthalmol 102: 584–587

Morgan SH, Maher ER, Purkiss P, Watts RWE, Curtis JR (1988) Oxalate metabolism in end-stage renal disease: the effect of ascorbic acid and pyridoxine. Nephrology Dialysis Transplantation (in press).

Morgan SH, Purkiss P, Watts RWE, Mansell MA (1987) Oxalate dynamics in chronic renal failure. Comparison with normal subjects and patients with primary hyperoxaluria. Nephron 47: 253 7

Nakatani T, Kawasaki Y, Minatogawa Y, Okuno E, Kido R (1985) Peroxisome localized human hepatic alanine: glyoxylate aminotransferase and its application to clinical diagnosis. Clin Biochem 18: 311–316

O'Callagan JW, Arbuckle SM, Craswell PW, Boyle PB, Searle JW, Smythe WR (1984) Rapid progression of oxalosis-induced cardiomyopathy despite adequate haemodialysis. Miner Electrolyte Metab 10: 48–51

Ratner S, Nocito V, Green DE (1944) Glycine oxidase. J Biol Chem 152: 119–133

Richardson KE, Farinelli MF (1981) The pathways of oxalate biosynthesis. In: Smith LH, Robertson WG, Finlayson B (eds) Urolithiasis, Clinical and Basic Research. Plenum Press, New York, pp 855–863

Robertson WG, Knowles F, Peacock M (1976) Urinary acid mucopolysaccharide inhibitors of calcium oxalate crystalisation. In: Fleisch H, Robertson WG, Smith LH, Vahlensieck W (eds) Urolithiasis Research. Plenum Press, New York, London, pp 331–338

Rose GA (1982) Urinary stones: clinical and laboratory aspects. MTP Press, Lancaster, pp 245–248

Rowsell EV, Carnie JA, Snell K, Taktak B (1972) Assays for glyoxylate aminotransferase activities. Int J Biochem 3: 247–257

Scheinman J, Najarian JS, Mauer SM (1984) Successful strategies for renal transplantation in primary oxalosis. Kidney Int 25: 804–811

Schlossberg MA, Bloom RJ, Richert DA, Westerfield WW (1970) Carboligase activity of α-ketoglutarate dehydrogenase. Biochemistry 9: 1148–1153

Scowen EF, Stansfeld AG, Watts RWE (1959) Oxalosis and primary hyperoxaluria. J Path Bact 77: 195–205

Vanrenterghem Y, Vandamme B, Lernt T, Michielsen P (1984) Severe vascular complications in oxalosis after successful cadaveric kidney transplantation. Transplantation 38: 93–95

Veall N, Gibbs DF (1982) The accurate determination of tracer clearance rates and equilibrium distribution volumes from single injection plasma measurements using numerical analysis. In: Joekes AM, Constable AR, Brown NJG, Tauxe WN (eds) Radionuclides in Nephrology. Academic Press/Gruene and Stratton, New York, pp 125–130

Wanders RJA, von Roermund CWT, Westra R, Schutgens RBH, van der Ender MA, Tager JM, Monnens LAH, Baadenhuysen H, Govaerts L, Przyrembel H, Wolff ED, Blom W, Huijmans JGM, van Laerhoven FGM (1987) Alanine:glyoxylate aminotransferase and the urinary excretion of oxalate and glycolate in hyperoxaluria type I and the Zellweger syndrome. Clin Chim Acta 165: 311–319

Watts RWE, Veall N, Purkiss P (1983) Sequential studies of oxalate dynamics in primary hyperoxaluria. Clin Sci 65: 627–633

Watts RWE, Veall N, Purkiss P (1984) Oxalate dynamics and removal rates during haemodialysis and peritoneal dialysis in patients with primary hyperoxaluria and severe renal failure. Clin Sci 66: 591–597

Watts RWE, Veall N, Purkiss P, Mansell MA, Haywood EF (1985a) The effect of pyridoxine on oxalate dynamics in three cases of primary hyperoxaluria (with glycollic aciduria). Clin Sci 69: 87–90

Watts RWE, Calne RY, Williams R, Mansell MA, Veall N, Purkiss P, Rolles K (1985b) Primary hyperoxaluria (Type I): attempted treatment by combined hepatic and renal transplantation. Q J Med 57: 697–703

Watts RWE, Calne RY, Rolles K, Danpure CJ, Morgan SH, Mansell MA, Williams R, Purkiss P (1987) Successful treatment of primary hyperoxaluria type I by combined hepatic and renal transplantation: correction of the enzymatic and metabolic defect. Lancet II: 474–475

Watts RWE, Morgan SH, Purkiss P, Mansell MA, Baker LRI, Brown CB (1988) Timing of renal transplantation in the management of pyridoxine resistant type I primary hyperoxaluria. Transplantation (in press)

Whelchel JD, Alison DV, Luke RG, Curtis J, Dietheim A (1983) Successful renal transplantation in hyperoxaluria. A report of two cases. Transplant 35: 161–164

Williams HE (1976) Oxalic acid: absorption, excretion and metabolism. In: Fleisch H, Robertson WG, Smith LH, Vahlensieck W (eds) Urolithiasis Research. Plenum Press, New York, London, pp 181–188

Williams HE, Smith LH Jr (1971) Possible pathogenic mechanism for hyperoxaluria in L-glyceric aciduria. Science 171: 390–391

Williams HE, Smith LH Jr (1983) Primary hyperoxaluria. In: Stanbury JB, Wyngaarden JB, Fredrickson DS, Goldstein JL, Brown MS (eds). The Metabolic Basis of Inherited Disease, 5 Edn. McGraw-Hill, New York, pp 204–228

Wise PJ, Danpure CJ, Jennings PR (1987) Immunological heterogeneity of hepatic alanine:glyoxylate aminotransferase in primary hyperoxaluria type I. FEBS Lett 222: 17–20

Yendt ER, Cohanim M (1986) Absorptive hyperoxaluria: A new clinical entity – successful treatment with hydrochlorothiazide. Clin Invest Med 9: 44–50

Chapter 6

Primary Hyperoxaluria in Children

T. M. Barratt, Vanessa von Sperling, M. J. Dillon, G. A. Rose and R. S. Trompeter

Introduction

Primary hyperoxaluria is an uncommon autosomal recessive disorder. The clinical features consist of nephrocalcinosis, recurrent renal stone, and progressive renal insufficiency, followed by the systemic deposition of oxalate crystals (oxalosis) (Williams and Smith 1983). In children the severity of the disorder ranges from death from renal failure in infancy to asymptomatic cases. The original anatomical description by Lepoutre in 1925 was of an infant dying of renal failure. The condition was first diagnosed in life by Newns and Black in 1953 in a girl of 12.5 yr with nephrocalcinosis and recurrent calcium-oxalate urolithiasis.

Type 1 hyperoxaluria results from a deficiency of the peroxisomal enzyme alanine:glyoxalate aminotransferase (Danpure and Jennings 1986; Danpure et al. 1987). Type 2 hyperoxaluria is characterised by L-glyceric aciduria (Williams and Smith 1968); it is a much less common disorder, but has recently been described in the UK in two children from unrelated families (Chalmers et al. 1984).

This review is based on the 16 children with primary hyperoxaluria seen at the Hospital for Sick Children, Great Ormond Street, and the Royal Free Hospital, London, from 1967 to 1987. The 11 cases seen at Great Ormond Street since 1980 comprise 6.7% of the 165 children with nephrocalcinosis and/ or renal calculi seen over the same period. In this population hyperoxaluria was diagnosed with about the same frequency as cystinuria and distal renal tubular acidosis.

Table 6.1. Primary hyperoxaluria in children 1967–1987

Patient	Sex	Presentation				Follow-up			
		Age (yr)	Plasma creatinine (μmol/l)	Urinary oxalate (mmol/1.73m²SA/24 h)	Urinary glycollate (mmol/1.73m²SA/24 h)	[e]Age (yr)	[f]B6 response	Plasma creatinine (μmol/l)	Current status
OA	M	0.2	553	—	—	0.3	0	257[d]	Dead
FT	F	0.3	172	—	—	0.5	0	283[d]	Dead
[a]RH	F	0.7	40	0.95	0.15	0.9	?	43	Well
[c]AA	M	1.7	1700	—	—	1.9	−	308[d]	Dead
[b]LC	F	1.7	62	2.22	1.69	5.7	+	432	Transplant
MW	F	1.9	140	1.42	0.61	3.3	0	60	Well
RS	F	3.5	48	1.12	0.94	6.9	++	60	Well
[b]GC	M	3.6	62	2.52	2.24	7.1	+	831	Continuous ambulatory peritoneal Dialysis
[c]FZ	M	4.6	708	—	—	4.7	−	629[d]	Dead
[a]DH	M	5.8	1300	0.22	0.05	6.0	−	432[d]	Continuous ambulatory peritoneal Dialysis
RD	M	6.6	57	3.02	3.25	10.9	0	132	Well
WH	M	8.6	134	1.13	—	14.0	+	120	Well
[c]GB	M	9.5	82	2.05	1.32	9.8	0	82	Well
BD	F	10.7	83	1.68	2.57	13.9	++→+	127	Well
AS	M	13.2	80	2.39	—	16.0	++	80	Well
AW	M	13.3	88	0.66	0.84	16.7	++	112	Well

[a,b]Siblings;
[c]Parental consanguinity;
[d]On dialysis;
[e]At follow-up or onset of end-stage renal failure;
[f]B6 response: ++, complete; +, partial; 0, none; −, not tested; ?, being tested.

Clinical Features

Age and Sex

Of the present series 10 were male and 6 female (Table 6.1). There were two pairs of siblings. Eight families were of Caucasian origin, the other 6 families coming from the Middle East and Indian subcontinent; parental consanguinity was noted in 3 of the latter group. Median age at presentation was significantly lower in the girls (1.8 yr) than in the boys (5.2 yr; $p < 0.05$, Mann-Whitney U-test), probably because of a greater susceptibility to urinary tract infection.

Presentation

Five cases presented in end-stage renal failure in early life, at 0.2, 0.3, 1.7, 4.6 and 5.8 years of age (Table 6.1). Case FT has been previously reported (Morris et al. 1982). Progression into renal failure was silent: all 5 cases required dialysis at presentation and did not recover renal function. In these children renal ultrasound showed crystallopathy (Fig. 6.1; Brennan et al. 1982), which was evident on the plain abdominal X-ray film (Fig. 6.2a) or computed tomography (CT) scan (Luers et al. 1980; Scheinman and Mahan 1986),

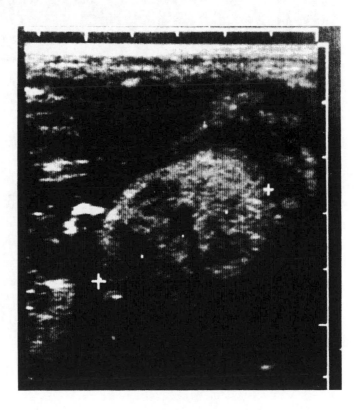

Fig. 6.1. Patient OA: renal ultrasound.

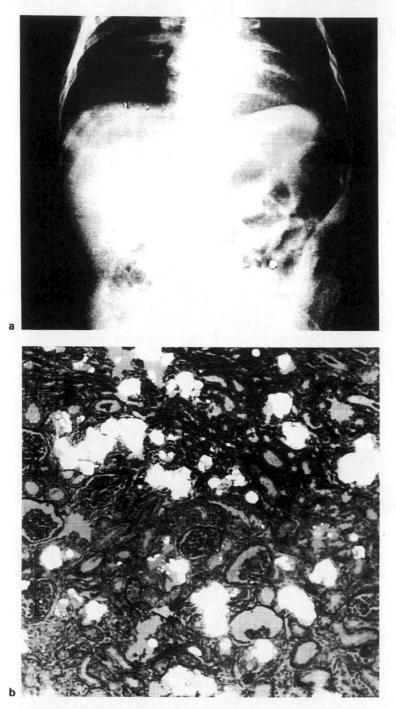

Fig. 6.2a,b.   Patient FT. **a** Plain abdominal X-ray film. **b** Renal biopsy under half-polarised light (courtesy Prof. RA Risdon).

distinguishing it from disorders of purine metabolism (HGPRTase and APRTase deficiency) which are the other principal causes of crystallopathy with renal failure of insidious onset in early childhood. The diagnosis was confirmed by open renal biopsy in FT (Fig. 6.2b), by bone-marrow examination in OA, AA, and FZ, and by enzyme analysis of a percutaneous liver-biopsy specimen in DH (Fig. 6.3a, Table 6.2). It is important to note that oxalate crystals were not evident in a routine marrow aspirate in AA but were clearly demonstrable in a trephine biopsy (Fig. 6.4). A pigmentary retinopathy was observed in OA (Fig. 6.5), as has been previously reported in other cases (Gottlieb and Ritter 1977; Fielder et al. 1980; Zak and Buncic 1983; Meredith et al. 1984).

Table 6.2. Family H

	GH	DH	RH
Age (yr)	7.0	5.8	0.7
Sex	M	M	F
Nephrocalcinosis	0	+++	++
Plasma creatinine (μmol/l)	38	1300	40
Plasma oxalate (μmol/l)	1.3	35.4	1.3
Urine oxalate (mmol/1.73m^{2}SA/24 h)	0.42	0.22	0.95
Urine glycollate (mmol/1.73m^{2}SA/24 h)	—	0.05	0.15
Hepatic [a]AGT (μmol/h/mg protein)	—	0.42	—
[b]GGT (μmol/h/mg protein)	—	0.77	—

[a]AGT: alanine:glyoxalate aminotransferase (normal 3.3–9.0 μmol/h/mg protein)
[b]GGT: glutamate:glyoxalate aminotransferase (normal 0.4–0.9 μmol/h/mg protein)
AGT and GGT measurements courtesy Dr. C.J. Danpure.

Infantile Oxalosis

This subset of patients has recently been reviewed by Leumann (1987) who found nearly 30 cases in the literature and added 2 of his own (Morris et al. 1982; Gilboa et al. 1983; de Zegher et al. 1984; Alinei et al. 1984; Leumann 1985). He pointed out that even though renal failure was severe and the renal parenchyma echodense, discrete renal calculi had not been described in these cases. It is not clear whether infantile oxalosis should be regarded as a separate entity or as the severe end of the spectrum of type 1 hyperoxaluria. The age of entry into end-stage renal failure in the present series does not have a bimodal distribution (Table 6.1), favouring the latter view. Danpure et al. (1987) reported an association between the clinical severity of the disorder in adults and the residual activity of hepatic alanine:glyoxalate aminotransferase, and in addition Danpure (1987, personal communication) has observed in one infant a low activity of the cytosolic enzyme glutamine:glyoxalate aminotransferase, which might account for the greater severity of the disturbance of glyoxalate metabolism in some younger children. These cases with early-onset renal failure should be distinguished from milder cases diagnosed in infancy, often as the result of family studies, eg. case RH (Table 6.2).

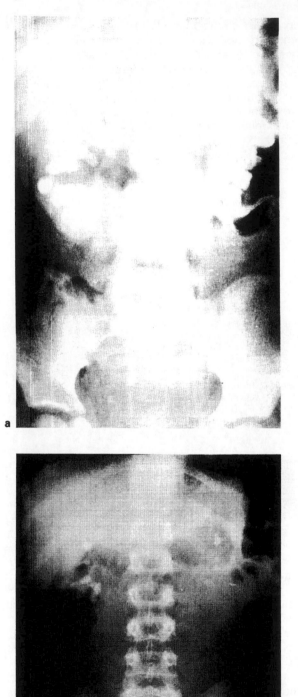

Fig. 6.3a,b. Plain
abdominal X-ray film.
a Anuric (patient DH).
b Moderately severe patient
RD).

Fig. 6.4. Patient AA: trephine bone marrow biopsy under half-polarised light (courtesy Prof RA Risdon).

Older Children

Later presentation is with urinary-tract infection or the passage of stone. The calculi may be held up at the ureterovesical junction with colic, or, distressingly, may impact at the urethral meatus. On X-ray films the typical finding is of nephrocalcinosis and renal calculi (Fig. 6.3). The principal differential diagnosis in children is distal renal tubular acidosis, and the two disorders may be confused as there is often a defect of urinary acidification in primary hyperoxaluria (Lagrue et al. 1959; Dent and Stamp 1970).

Biochemical Features

Urine Oxalate in Normal Children

Data on 24-h oxalate excretion in normal children are very scanty (Table 6.3), and are virtually nonexistent for children under 2 yr of age, nor are there any using modern analytical methods (Hallson and Rose 1974; Kasidas and Rose 1985). Gibbs and Watts (1969) reported that in healthy children aged 2–14 yr oxalate excretion corrected to 1.73 m^2 body surface area (SA) was the same

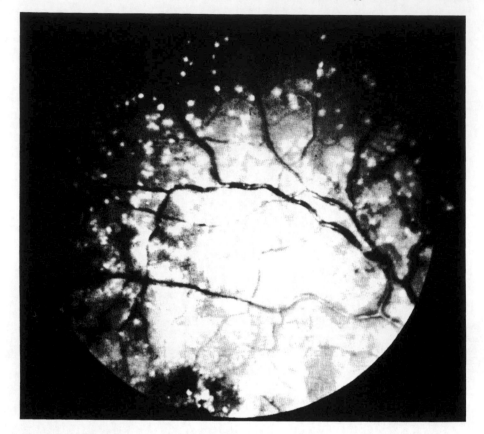

Fig. 6.5.   Patient OA: retinopathy (courtesy Mr J Bolger).

Table 6.3.   Urinary oxalate excretion in healthy children

Authors (Method)	Children studied			Urine oxalate	
	Sex	Age (yr)	Number	(mmol/1.73m²SA/24 h) Mean	Range (±2SD)
Hockaday et al. 1965 (Isotope dilution)	Both	3–13	15	0.37	0.11–0.62
Gibbs and Watts 1969 (Isotope dilution)	M	3–14	8	0.45	0.33–0.53
	F	2–14	7	0.42	0.31–0.63
Hodgkinson and Williams 1971 (Colorimetric)	M	4–12	4	0.39	0.31–0.49
	F	3–13	8	0.36	0.20–0.69
Ogilvie et al. 1976 (Colorimetric)	Both	0.3–17	16	0.27	0.16–0.37

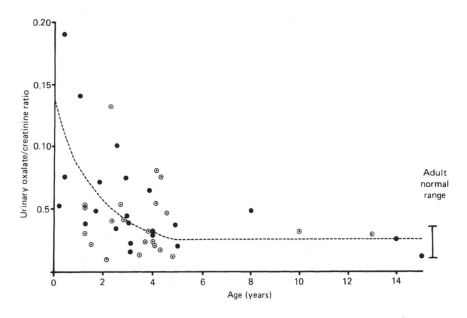

Fig. 6.6. Urine oxalate/creatinine ratio in normal children. Filled circles, males; open circles, females (Kasidas and Rose 1987).

as in adults, but inspection of their data suggests that the level thus corrected rises during childhood. Using an enzymatic method the normal oxalate excretion in adults is 0.30 ± 0.08 (SD) mmol/24h (1 mmol = 90 mg anhydrous oxalic acid) (Hallson and Rose 1974). In the absence of better data the upper limit of urinary oxalate excretion in healthy children is currently taken to be 0.46 mmol/1.73m²SA/24h.

The normal urinary oxalate/creatinine ratio is high in young children, especially those under 2 years of age, and declines rapidly (Gibbs and Watts 1969; Kasidas and Rose 1987; Fig. 6.6). This relationship with age is to be expected if oxalate excretion is proportional to SA (and thus to the square of height) since creatinine excretion is proportional to body weight (and thus to the cube of height): the oxalate/creatinine ratio in healthy children would then be inversely proportional to height and hence would decline fourfold from infancy to adult life.

Hyperoxaluria

Given these uncertainties in the normal reference ranges, it is fortunate that urinary oxalate excretion in children with primary hyperoxaluria is substantially above the upper limit of normal, leaving little room for doubt about the diagnosis in most cases. In the 11 children without renal failure (plasma creatinine <150 μmol/l) in the present series urine oxalate excretion was 1.74 ± 0.75 (SD) mmol/1.73 m²SA/24 h (Table 6.1). Nevertheless, in young children considerable care is needed to establish the diagnosis irrefutably. Other causes of hyperoxaluria must be considered: gastrointestinal disease results in quite substantial increases in oxalate excretion in some children, particularly in those with ileal resection or pancreatic disease (Ogilvie et al. 1976), and modest increases in oxalate excretion may also be the consequence of a high dietary intake of oxalate though this effect is poorly documented in children.

Glycollate

Urinary excretion of glycollate is increased in primary hyperoxaluria. Using an isotope dilution method Hockaday et al. (1965) reported urinary excretion of glycollate in 15 normal children aged 3.3–13.5 yr to be 0.55 mmol/1.73 m^{2}SA/24 h with an upper limit of 0.90. However, using a specific enzymatic method the normal urinary excretion in adults is 0.19 ± 0.07 mmol/24 h (Kasidas and Rose 1979). There are no comparable data for children, nor is it known whether surface area correction is appropriate. In the 9 hyperoxaluric children without renal failure in the present series in whom the data were available glycollate excretion was 1.51 ± 1.01 mmol/1.73 m^{2}SA/24 h. The mean molar glycollate/oxalate ratio in these children was 85 ± 42%, and increased strikingly with age (r = 0.76, p<0.05; Fig. 6.7).

Diagnosis in Renal Failure

Diagnostic problems arise in the presence of renal failure when oxalate excretion declines to within the normal range (as in case DH), and ceases

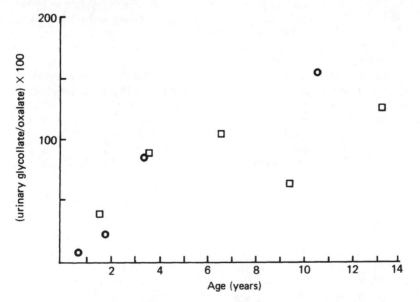

Fig. 6.7. Relationship between urinary glycollate/oxalate ratio and age in children with primary hyperoxaluria and relatively well-preserved renal function (plasma creatinine <150 μmol/l: r=0.76). Squares, males; circles, females.

altogether when anuria supervenes. The concentration of oxalate in plasma is raised in renal failure, and although it tends to be higher for any given plasma creatinine concentration in primary hyperoxaluria than in other forms of renal failure (Constable et al. 1979), the overlap is large and the relationship between oxalate and creatinine concentrations in plasma at different levels of renal function has not been established for children. In these circumstances the diagnosis depends upon the detection of systemic deposits of oxalate or, better, the demonstration of the specific enzyme defect in liver tissue (Danpure et al. 1987; Table 6.2).

Progress

Renal Function

The 5 children in the present series who presented in renal failure did not recover function: 3 are dead, one has been transplanted (though with oxalate deposits in the graft), and the fifth is on continuous ambulatory peritoneal dialysis (Table 6.1). Two other children declined into end-stage renal failure, 3 have had a significant increase in the concentration of creatinine in the plasma over periods of observation from 3 to 5 yr, 4 appear to be stable, and 2 have only been followed up for periods of less than one year (Fig. 6.8). The longterm outlook for renal function is without doubt poor, but with increasing awareness of the condition milder cases are being recognised (Gill and Rose 1986: See also chapter 8) and the prognosis of currently diagnosed cases is not necessarily the same as that of earlier series.

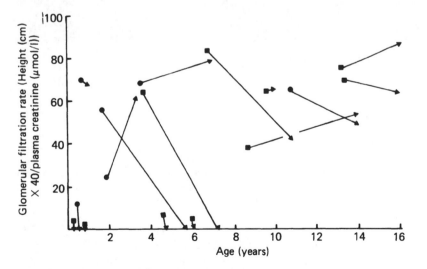

Fig. 6.8. Evolution of glomerular filtration rate, estimated from the plasma creatinine concentration and the height. Filled squares, males; filled circles, females. (Counahan et al. 1976).

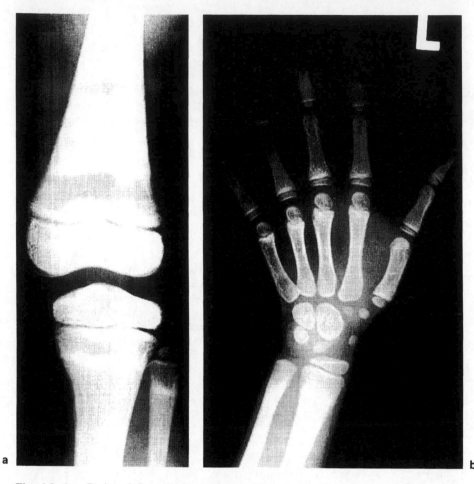

Fig. 6.9a,b.   Patient LC: oxalosis of bone. **a** Knee. **b** Hand.

Systemic Oxalosis

With advancing renal failure, plasma oxalate concentration rises and oxalate crystals are deposited at extrarenal sites. A crippling osteopathy may ensue, compounded by uraemic osteodystrophy (Breed et al. 1981; Wiggelinkhuizen and Fisher 1982; Fig. 6.9). The bony lesions are sometimes, but not always, reversed by successful renal transplantation (Scheinman et al. 1984). Cardiac deposits may cause conduction defects (Tonkin et al. 1976), and a cardiomyopathy has been described in a four-year-old boy (Pikula et al. 1973). Arterial lesions may cause limb gangrene (Arbus and Sniderman 1974; Blackburn et al. 1975), though arterial spasm may be reversible (Van Damme-Lombaerts et al. 1988). The severity of these complications is such that, if renal transplantation is to be effective, it is best undertaken early in the progression of the renal failure.

Treatment

Simple Measures

The maintenance of hydration and the prompt recognition and treatment of infections of the urinary tract are very important: renal function in both LC and GC declined substantially and irreversibly following an episode of dehydration associated with urinary-tract infection. Magnesium supplements have been recommended as magnesium ions inhibit the precipitation of calcium oxalate (Dent and Stamp 1970), but it is difficult to influence urinary magnesium excretion in children without causing diarrhoea, and magnesium supplements should not be used if renal failure has supervened. Some children with primary hyperoxaluria are also hypercalciuric (Proesmans 1987, personal communication), and phosphate supplements and/or a thiazide diuretic may then be appropriate. Restraint should certainly be exercised with vitamin-D-analogue therapy. If there is a significant acidification defect alkali therapy is indicated.

Pyridoxine

Some patients with type 1 primary hyperoxaluria respond to pyridoxine (Gibbs and Watts 1967, 1970; Kasidas and Rose 1984; Table 6.4). Occasionally there is complete restoration of urine oxalate excretion to within the normal range, as in RS (Fig. 6.10a), but more often the response is partial with a significantly reduced but nevertheless still abnormal oxalate excretion (WH; Fig. 6.10b). Large doses of pyridoxine were used in earlier studies, but it is recognised that some patients may respond to doses as small as 10 mg daily (Yendt and

Table 6.4. Urinary oxalate excretion: response to pyridoxine

Patient	Age (yr)	[a]Urinary oxalate excretion at a range of B6 levels (dosage mg/24 h)				[b]Response to pyridoxine
		Pre-B6	<20	20–100	>100	
RS	3.5	1.12	0.47	0.39		++
AW	13.3	0.62			0.27	++
BD	10.7	3.56	0.34	0.43	1.92	++→+
LC	1.7	2.22			1.42	+
GC	3.6	2.52			1.89	+
WH	8.6	1.72			1.02	+
AS	13.2	2.39			1.58	+
MW	1.9	1.42	1.96		1.40	0
RD	6.6	3.02	3.09		5.94	0
GB	9.5	2.44	2.67		3.29	0

[a]Urinary oxalate excretion (mmol/1.73m^{2}SA/24 h): data given are the last observation before the start of B6 therapy and the first observation 6 weeks after the start of the stated dose
[b]B6 response: ++, complete; +, partial; 0, none

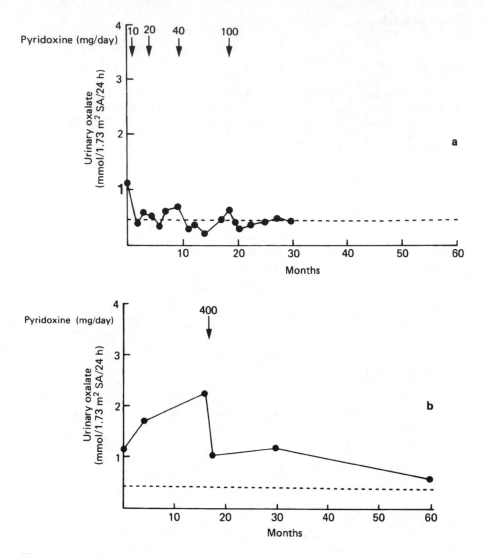

Fig. 6.10a–c. Response to pyridoxine. **a** Complete response (patient RS: 3.5 yr). **b** Partial response (patient WH: 8.6 yr). **c** Transient response (patient BD: 10.6 yr). The doses of pyridoxine were started where indicated by the arrows.

Cohanim 1985; Gill and Rose 1986). Case BD tantalisingly responded to 10 mg pyridoxine daily with complete normalisation of oxalate excretion, but after a few months oxalate excretion rose again in spite of a pyridoxine dosage increased to 100 and then to 400 mg daily (and compliance confirmed by measurement of 4-pyridoxic acid excretion) (Fig. 6.10c). After each rise in dose there was a fall in urinary oxalate that was not maintained. A similar unsustained response has been described in an adult patient (Gill and Rose 1986). Three infants have been reported in whom successful treatment with

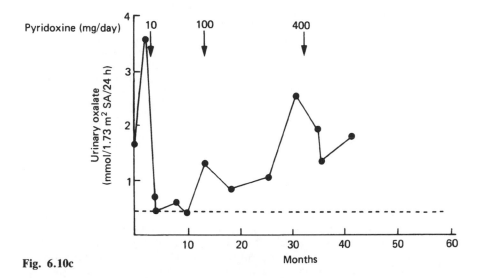

Fig. 6.10c

pyridoxine was started between one and two months of age: two of them came from families in which other siblings had died of renal failure with oxalosis in infancy (Rose et al. 1982; Alinei et al. 1984; de Zegher et al. 1985). Pyridoxine therapy was ineffective in 2 patients with infantile oxalosis in the present series (OA, FT), and was of no value once renal failure was established. Neurological complications have been attributed to very high doses of pyridoxine (Schaumburg et al. 1983), and reported in a 10-week-old infant receiving 1000 mg daily (de Zegher et al. 1985). In view of the reports of response to low doses and the potential toxicity of larger doses, trials of pyridoxine therapy in children with primary hyperoxaluria should proceed stepwise from a low dose with at least two months of observation at each dosage level (except in infants in whom establishment of effective therapy is more urgent).

Surgery and Lithotripsy

Surgical intervention may be required if the stones are obstructive and show no signs of passing spontaneously. Modern methods of stone removal are appropriate, but caution should be exercised with extracorporeal shock-wave lithotripsy: the two cases that were treated (RD and GC) both showed significant deterioration of renal function on the treated side (Boddy et al. 1988; Table 6.5). The 8-year-old boy graphically expressed his views of this form of treatment (Fig. 6.11).

Dialysis and Renal Transplantation

Treatment of end-stage renal failure by dialysis is unsatisfactory, as systemic oxalosis inevitably progresses (Watts et al. 1984). Haemodialysis is more

Table 6.5. Extracorporeal shock wave lithotripsy (ESWL)

Patient	Treatment	Renal function	Pre-treatment	Post-treatment
GC	ESWL ×6 left kidney	Plasma creatinine (μmol/l)	140	300
		^{51}Cr-EDTA GFR (ml/min/1.73m^{2}SA)	32	16
		^{99m}Tc-DMSA uptake %: left	46	41
		right	54	59
RD	ESWL ×2 left kidney	Plasma creatinine (μmol/l)	105	145
		^{51}Cr-EDTA GFR (ml/min/1.73m^{2}SA)	39	33
		^{99m}Tc-DMSA uptake %: left	44	34
		right	56	66

effective than peritoneal dialysis in removing oxalate (Watts et al. 1984), but vascular complications may limit access (Arbus and Sniderman 1974). Recurrence of oxalate deposits with allograft failure has raised the question as to whether any form of treatment of end-stage renal failure is appropriate for children with oxalosis (Donckerwolcke et al. 1980). However, prolonged allograft survival has been described (Leumann et al. 1978), and Scheinman et al. (1984) reported excellent results in 3 of 6 children aged 0.5–8 yr treated by parent-donor renal transplantation with aggressive perioperative haemodialysis and post-transplant therapy with pyridoxine, magnesium salts, neutral phosphate supplements, thiazide diuretics, and copious hydration.

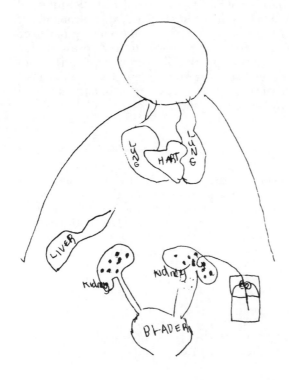

Fig. 6.11. Body image of patient GC (aged 7.5 yr) following extracorporeal shock wave lithotripsy.

These results have engendered a new optimism in the treatment of end-stage renal failure in children with primary hyperoxaluria, but it is difficult to imagine that in the long run the disease will not recur in the graft unless the patient is one of the fortunate few who are fully and permanently responsive to pyridoxine.

The Future

Primary hyperoxaluria remains a distressing and serious disorder for which the ultimate goal is prevention, or, failing that, correction of the metabolic defect.

Antenatal Diagnosis

There is conflicting evidence as to whether primary hyperoxaluria is reflected in raised oxalate concentration in amniotic fluid (Rose et al. 1982; Leumann et al. 1986). Now that the primary enzyme defect has been identified, antenatal diagnosis by foetal liver biopsy is theoretically feasible, and no doubt gene probes will be forthcoming in the future (Danpure 1986).

Liver Transplantation

The primary defect is hepatic, and liver transplantation has been shown to correct the metabolic error in an adult with primary hyperoxaluria (Watts et al. 1985). With the improving results of the paediatric liver-transplant programmes it is likely that in the future children severely affected by this disease will be treated by combined liver–kidney transplantation.

Acknowledgement. We thank Sister J. Fay for data retrieval.

References

Alinei P, Guignard JP, Jaeger P (1984) Pyridoxine treatment of type 1 hyperoxaluria. N Engl J Med 311: 798–799
Arbus GS, Sniderman S (1974) Oxalosis with peripheral gangrene. Arch Pathol 97: 101–110
Blackburn WE, McRoberts JW, Bhathena D, Vazquez M, Luke RG (1975) Severe vascular complications in oxalosis after bilateral nephrectomy.Ann Int Med 82: 44–46
Boddy SM, Duffy PG, Barratt TM, Whitfield HN (1988) Hyperoxaluria and renal calculi in children: the role of extracorporeal shock wave lithotripsy. Proc Roy Soc Med, in press
Breed A, Chesney R, Friedman A, Gilbert E, Langer L, Lattoraca R (1981) Oxalosis-induced bone disease: a complication of transplantation and prolonged survival in primary hyperoxaluria. J Bone Joint Surg 63: 310–316
Brennan JN, Diwan RV, Makker SP, Cromer BA, Bellon EM (1982) Ultrasonic diagnosis of primary hyperoxaluria in infancy. Radiology 145: 147–148

Chalmers RA, Tracey BM, Mistry J, Griffiths MD, Green A, Winterborn MH (1984) L-glyceric aciduria (primary hyperoxaluria type 2) in siblings in two unrelated families. J Inherited Metab Dis 7: 133–134

Constable AR, Joekes AM, Kasidas GP, Rose GA (1979) Plasma level and clearance of oxalate in normal subjects and in patients with primary hyperoxaluria or chronic renal failure. Clin Sci 56: 299–304

Counahan R, Chantler C, Ghazali S, Kirkwood B, Rose F, Barratt TM (1976) Estimation of glomerular filtration rate from plasma creatinine concentration in children. Arch Dis Child 51: 875–878

Danpure CJ (1986) Peroxisomal alanine glyoxalate aminotransferase and prenatal diagnosis of primary hyperoxaluria type 1. Lancet II: 1168

Danpure CJ, Jennings PR (1986) Peroxisomal alanine:glyoxalate aminotransferase deficiency in primary hyperoxaluria type 1. FEBS Lett 201: 20–24

Danpure CJ, Jennings PR, Watts RWE (1987) Enzymological diagnosis of primary hyperoxaluria type 1 by measurement of hepatic alanine:glyoxylate aminotransferase activity. Lancet I: 289–291

Dent CE, Stamp TCB (1970) Treatment of primary hyperoxaluria. Arch Dis Child 45: 735–745

Donckerwolcke RA et al. (1980) Combined report on regular dialysis and transplantation of children in Europe, 1979. Proc Eur Dial Transplant Assoc 17: 87–115

Fielder AR, Garner A, Chambers TL (1980) Ophthalmic manifestations of primary oxalosis. Br J Ophthalmol 64: 782–788

Gibbs DA, Watts RWE (1967) Biochemical studies on the treatment of primary hyperoxaluria. Arch Dis Child 42: 505–508

Gibbs DA, Watts RWE (1969) The variation of urinary oxalate excretion with age. J Lab Clin Med 73: 901–908

Gibbs DA, Watts RWE (1970) The action of pyridoxine in primary hyperoxaluria. Clin Sci 38: 277–286

Gilboa N, Largent JA, Urizar RE (1983) Primary oxalosis presenting as anuric renal failure in infancy: diagnosis by X-ray diffraction of kidney tissue. J Pediatr 103: 88–90

Gill HS, Rose GA (1986) Mild metabolic hyperoxaluria and its response to pyridoxine. Urol Int 41: 393–396

Gottlieb RP, Ritter JA (1977) "Flecked retina" – an association with primary hyperoxaluria. J Pediatr 90: 782–788

Hallson PC, Rose GA (1974) A simplified enzymatic method for determination of urinary oxalate. Clin Chim Acta 55: 29–39

Hockaday TDR, Frederick EW, Clayton JE, Smith LH (1965) Studies on primary hyperoxaluria. II. Urinary oxalate, glycollate and glyoxalate measurement by isotope dilution methods. J Lab Clin Med 65: 667–687

Hodgkinson A, Williams A (1972) An improved colorimetric procedure for urine oxalate. Clin Chim Acta 36: 127–132

Kasidas GP, Rose GA (1979) A new enzymatic method for the determination of glycollate in urine and plasma. Clin Chim Acta 96: 25–36

Kasidas GP, Rose GA (1984) Metabolic hyperoxaluria and its response to pyridoxine. In: Ryall R, Brockis JG, Marshall V, Finlayson B (eds) Urinary stone. Churchill Livingstone, Melbourne, pp 138–147

Kasidas GP, Rose GA (1985) Continuous-flow assay for urinary oxalate using immobilised oxalate oxidase. Ann Clin Biochem 22: 412–419

Kasidas GP, Rose GA (1987) The measurement of plasma oxalate and when this is useful. In: Vahlensieck W, Gasser G (eds) Pathogenese und klinik der harnsteine 12. Steinkopff, Darmstadt, pp 140–147

Lagrue G, Laudat MH, Meyer P, Sapir M, Milliez P (1959) Oxalose familiale avec acidose hyperchlorémique secondaire. Sem Hop Paris 35: 2023–2032

Lepoutre C (1925) Calculs multiples chez l'enfant. Infiltration du parenchyme rénale par des cristaux. J Urol méd chir 20: 424–429

Leumann EP (1985) Primary hyperoxaluria: an important cause of renal failure in infancy. Int J Paediatr Nephrol 6: 13–16

Leumann EP (1987) New aspects of infantile oxalosis. Pediatr Nephrol 1: 531–535

Leumann E, Matasovic A, Niederwieser A (1986) Primary hyperoxaluria type 1: oxalate and glycolate unsuitable for prenatal diagnosis. Lancet II: 340

Leumann EP, Wegmann W, Largiader F (1978) Prolonged survival after renal transplantation in primary hyperoxaluria of childhood. Clin Nephrol 9: 29–34

Luers PR, Lester PD, Siegler RL (1980) CT demonstration of cortical nephrocalcinosis in congenital oxalosis. Pediatr Radiol 10: 116–118

Meredith TA, Wright JD, Gammon JA, Fellner SK, Warshaw BL, Maio M (1984) Ocular involvement in primary hyperoxaluria. Arch Ophthalmol 102: 584–587

Morris MC, Chambers TL, Evans PW, Malleson PN, Pincott JR, Rose GA (1982) Oxalosis in infancy. Arch Dis Child 57: 224–228

Newns GH, Black JA (1953) A case of calcium oxalate nephrocalcinosis. Great Ormond Street Journal 5: 40–44

Ogilvie D et al. (1976) Urinary outputs of oxalate, calcium, and magnesium in children with intestinal disorders. Arch Dis Child 51: 790–795

Pikula B, Plamenac P, Circic B, Nikulin A (1973) Myocarditis caused by primary oxalosis in a 4 year old child. Virchows Arch (Pathol Anat) 358: 99–103

Rose GA, Arthur LJ, Chambers TL, Kasidas GP, Scott IV (1982) Successful treatment of primary hyperoxaluria in a neonate. Lancet I: 1298–1299

Schaumburg H, Kaplan J, Windebank A, Vick N, Rasmus S, Pleasure D, Brown MJ (1983) Sensory neuropathy from pyridoxine abuse. N Engl J Med 309: 445–448

Scheinman JI, Mahan J (1986) Radiological aspects of primary hyperoxaluria. Am J Radiol 146: 395–401

Scheinman JI, Najarian JS, Mauer SM (1984) Successful strategies for renal transplantation in primary oxalosis. Kidney Int 25: 804–811

Tonkin AM, Mond HG, Mathew TH, Sloman JG (1976) Primary oxalosis with myocardial involvement and heart block. Med J Aust 1: 873–874

Van Damme-Lombaerts R, Proesmans W, Alexandre G, Van Dyck M, Serus M, Wilms G (1988) Reversible arterial spasm in an adolescent with primary oxalosis. Clin Nephrol, in press

Watts RWE, Veall N, Purkiss P (1984) Oxalate dynamics and removal rates during haemodialysis and peritoneal dialysis in patients with primary hyperoxaluria and severe renal failure. Clin Sci 66: 591–597

Watts RWE et al. (1985) Primary hyperoxaluria (type 1): attempted treatment by combined hepatic and renal transplantation. Q J Med 57: 697–703

Wiggelinkhuizen J, Fisher RM (1982) Oxalosis of bone. Pediatr Radiol 12: 307–309

Williams HE, Smith LH (1968) L-Glyceric aciduria: a new genetic variant of primary oxaluria. N Eng J Med 278: 233–239

Williams HE, Smith LH (1983) Primary hyperoxaluria. In: Stanbury JB, Wyngaarden JB, Fredrickson DS, Goldsteiin JL, Brown MS (eds) The metabolic basis of inherited disease. McGraw-Hill, New York, 3rd ed, pp 204–228

Yendt ER, Cohanim M (1985) Response to a physiologic dose of pyridoxine in type 1 primary hyperoxaluria (1985) N Engl J Med 312: 953–957

Zak TA, Buncic R (1983) Primary hereditary oxalosis retinopathy (1983) Arch Ophthalmol 101: 78–80

de Zegher FE, Wolff ED, Heijden AJ, Sukhai RN (1984) Oxalosis in infancy. Clin Nephrol 22: 114–120

de Zegher FE, Przyrembel H, Chalmers RA, Wolff ED, Huijmans JGM (1985) Successful treatment of infantile type 1 primary hyperoxaluria complicated by pyridoxine toxicity. Lancet II: 392–393

Chapter 7

Enteric and Other Secondary Hyperoxalurias

D. S. Rampton and M. Sarner

Introduction

It has been known for 25 years that nephrolithiasis occurs more commonly in patients with intestinal disease than in the general population (Deren et al. 1962; Gelzayd et al. 1968). The majority of stones in these patients are composed predominantly of calcium oxalate (Gelzayd et al. 1968), and a mechanism to explain this observation became apparent with the demonstration about 15 years ago that individuals with ileal disease or resection excrete increased amounts of oxalate in the urine (Dowling et al. 1971; Admirand et al. 1971; Smith et al. 1972). The aim of this paper is to review present knowledge about the cause, clinical significance and treatment of enteric hyperoxaluria (EHO). In the course of a resumé of pertinent aspects of oxalate metabolism, other types of secondary hyperoxaluria will be briefly mentioned.

Oxalate Metabolism and Secondary Hyperoxaluria

Oxalate is excreted mainly in the urine as an end-product of metabolism (Hodgkinson 1977; Earnest 1979; Laker 1983). Under normal circumstances urinary oxalate is derived principally from glyoxylate, ascorbic acid and dietary oxalate and the relevant metabolic pathways are outlined in Fig. 7.1. Primary hyperoxaluria is discussed in Chapters 5 and 6 and hyperoxaluria due to pyridoxine deficiency in Chapter 11. Other causes of secondary hyperoxaluria are listed in Table 7.1.

Rarely, renal oxalosis and/or hyperoxaluria may be due to excessive hepatic synthesis of oxalate from ethylene glycol, an antifreeze agent (Parry and

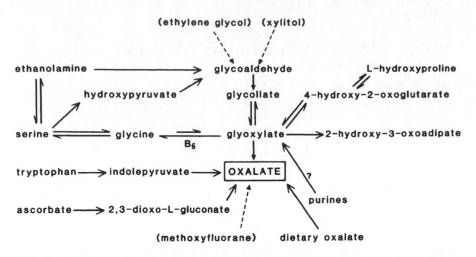

Fig. 7.1. Summary of oxalate metabolism in man. Compounds shown in parentheses are rare exogenous cause of hyperoxaluria (see Table 7.1, below).

Wallach 1974), methoxyfluorane, a general anaesthetic (Frascino et al. 1970) and xylitol, used in some parenteral nutrition regimes (Thomas et al. 1972; Rofe et al. 1979). Hyperoxaluria has been reported in patients receiving bladder irrigation with glycine-rich fluids after prostatectomy (Fitzpatrick et al. 1981) and also occurs after intake of purines (Zarembski and Hodgkinson 1969, Simmonds et al. 1981) and ascorbic acid in excess (Briggs et al. 1973; Schmidt et al. 1981). Exogenous oxalate poisoning with fatal outcome has been described in patients taking the compound orally (Jeghers and Murphy 1945) and given it intravenously (Dvorackova 1966). Much the commonest cause of secondary hyperoxaluria, however, is enteric hyperoxaluria in which, as will be shown,

Table 7.1. Causes of secondary hyperoxaluria

	References
1. *Increased endogenous synthesis of oxalate*	
Pyridoxine deficiency	See Chap. 11
2. *Excessive intake of oxalate precursors*	
Ascorbic acid	Briggs et al. 1973, Schmidt et al. 1981
Ethylene glycol	Parry and Wallach 1974
Methoxyfluorane	Frascino et al. 1970
Xylitol	Thomas et al. 1972, Rofe et al. 1979
Glycine	Fitzpatrick et al. 1981
Purines	Zarembski and Hodgkinson 1969; Simmonds et al. 1981
3. *Increased intake/absorption of oxalate*	
Oxalate overdose	Jeghers and Murphy 1945; Dvorackova 1966
Enteric hyperoxaluria	See text

hyperabsorption of dietary oxalate occurs in patients with a variety of digestive and absorptive bowel disorders characterised by malabsorption of fat, bile salts, or both (Earnest 1979): these disorders include ileal disease or resection (Dowling et al. 1971; Admirand et al. 1971; Smith et al. 1972), jejuno-ileal bypass (Dickstein and Frame 1973; Gregory et al. 1977; Stauffer 1977a; Hofmann et al. 1981, 1983), coeliac disease (McDonald et al. 1977) and pancreatic insufficiency (Andersson and Gillberg 1977; Dobbins and Binder 1977; Stauffer 1977b; Rampton et al. 1979).

Pathogenesis of Enteric Hyperoxaluria

Three main theories have been devisd to explain the hyperoxaluria observed in patients with intestinal disease, but the first has failed to withstand critical experimental evaluation.

Glyoxylate Theory

According to this early ingenious proposal, failure of absorption of glycine-conjugated bile acids by the resected or diseased terminal ileum leads first to deconjugation and then conversion of the liberated glycine, by colonic bacterial flora, to glyoxylate: the latter, it was hypothesized, is then absorbed and oxidised in the liver to oxalate (Hofmann et al. 1970).

Three lines of evidence, however, militate against this proposal. First, although increased formation of glyoxylate would be expected to result in enhanced urinary excretion of glycollate as well as oxalate, several studies have shown that output of the former is normal in patients with EHO (Admirand et al. 1971; Earnest et al. 1974; Rampton et al. 1979). Second, oral administration of ^{14}C-labelled glycocholic acid to patients with ileal resection was not, in the event, followed by excretion in the urine of ^{14}C-labelled oxalate (Chadwick et al. 1973; Hofmann et al. 1973). Third, oral ingestion of taurine by such patients, which reduces the ratio of glycine- to taurine-conjugated bile acids, does not consistently reduce urinary oxalate excretion (Dowling et al. 1971; Admirand et al. 1971; Smith et al. 1972; Chadwick et al. 1973). Other explanations for EHO were therefore needed.

Origin of Excess Urinary Oxalate

It is now clear that the increased amounts of oxalate excreted in the urine of patients with EHO are derived principally from food. Thus, while normal subjects absorb 5%–10% of ingested oxalate, patients with EHO absorb up to 60% of oral doses of ^{14}C-labelled oxalate (Chadwick et al. 1973; Earnest et al. 1974; Caspary et al. 1977; Hylander et al. 1978; Tiselius et al. 1981; Hofmann et al. 1983) and dietary loads of stable oxalate (Stauffer et al. 1973; Earnest et al. 1974; Barilla et al. 1978; Rampton et al. 1979; Hofmann et al.

1983). While the restoration to normal values of urinary excretion of oxalate produced by adherence to a low-oxalate diet confirms its dietary origin in most patients with EHO (Chadwick et al. 1973; Stauffer et al. 1973; Earnest et al. 1974; Hofmann et al. 1983), in one study hyperoxaluria persisted in subjects with a jejuno-ileal bypass despite a very low intake of oxalate (Hofmann et al. 1983). The reduction in urinary oxalate which followed deprivation of dietary protein as well as oxalate suggested that, in this small group of patients, urinary oxalate may be partly derived from tissue (Ribaya and Gershoff 1982) or bacterial synthesis, either of oxalate itself, or of an oxalate precursor, from dietary protein (Hofmann et al. 1983).

Reasons for Hyperabsorption of Dietary Oxalate

Hyperabsorption of dietary oxalate in patients with EHO may be related to changes in oxalate solubility within the gut lumen (the "solubility" theory) or to bile salt and fatty acid-induced increases in colonic mucosal permeability (the "permeability theory"), but the two mechanisms are not mutually exclusive.

Solubility Theory

Evidence adduced both in vivo and in vitro supports the proposal that, in patients with steatorrhoea, malabsorbed fatty acids form soaps with intraluminal calcium, thereby increasing the amount of soluble oxalate available for absorption (Fig. 7.2) (Andersson and Jagenburg 1974; Binder 1974; Earnest et al. 1974; McDonald et al. 1977).

First, a positive linear correlation between urinary excretion of oxalate and the degree of steatorrhoea, whatever the intestinal pathology, has been demonstrated repeatedly (Fig. 7.3) (Andersson and Jagenburg 1974; Earnest et al. 1974; Andersson and Gillberg 1977; McDonald et al. 1978; Stauffer 1977a,b; Andersson et al. 1978; Hylander et al. 1978; Rampton et al. 1979, 1984). More importantly, urinary excretion of oxalate can be increased or reduced in patients with EHO by parallel changes in the dietary intake of fat (Andersson and Jagenburg 1974; Earnest et al. 1974, 1975a,b; Andersson et al. 1978). Furthermore, as would be predicted by the solubility theory, addition of calcium to an otherwise constant diet leads to a fall in the absorption and

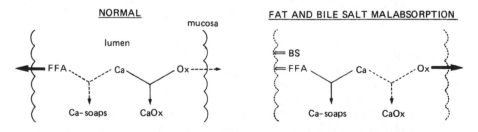

Fig. 7.2. Pathogenesis of colonic hyperabsorption of oxalate in patients with malabsorption of fat and bile salts according to the solubility and permeability theories (FFA, free fatty acid; Ca, calcium; Ox, oxalate; BS, bile salts).

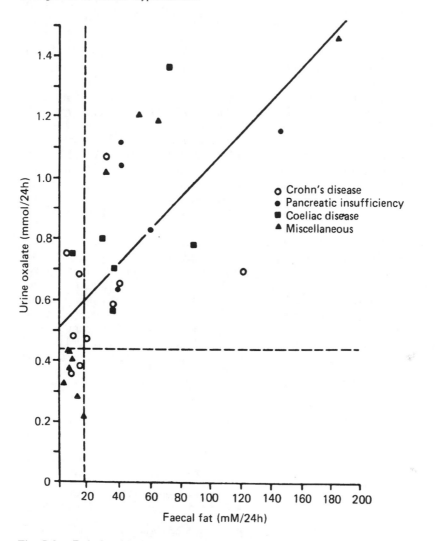

Fig. 7.3. Relationship between 24-h urinary oxalate and faecal fat output in patients on a high oxalate intake (r = 0.73, p<0.001). The dashed lines indicate the upper limit of oxalate excretion (horizontal) and of faecal fat output (vertical) in normal subjects. (Reproduced with permission from Rampton et al. 1979).

urinary excretion of oxalate (Earnest et al. 1975a,b; Caspary et al. 1977; Stauffer 1977a,b; Barilla et al. 1978; Hylander et al. 1980). In vitro experiments have confirmed that when sodium oleate is present in excess, oxalate remains in solution despite addition of calcium chloride, and that, since addition of calcium oxalate to a micellar solution of sodium oleate evokes precipitation of calcium oleate, the affinity of calcium for oleate is greater than for oxalate (Binder 1974; Earnest et al. 1975b). How closely these test-tube results, however, reflect events in the far more complex chemical environments of the intestinal lumen and near the mucosal surface is uncertain.

Permeability Theory

In many patients with malabsorption syndromes, and particularly in those with ileal disease or resections, there is spillover into the colon of bile salts and fatty acids (Hofmann and Poley 1972). Both produce a dose-related stimulation of colonic secretion of water and electrolytes and, at least in animal models, histological evidence of mucosal damage (Mekhjian and Phillips 1970; Mekhjian et al. 1971; Binder 1973; Rampton et al. 1981). There are now abundant data, obtained using a variety of techniques in rats (Chadwick et al. 1975; Saunders et al. 1975; Dobbins and Binder 1976; Caspary 1977; Schwartz et al. 1980; Kathpalia et al. 1984), monkeys (Chadwick et al. 1975) and man (Fairclough et al. 1977), indicating that, in addition, bile salts and fatty acids increase colonic mucosal permeability to oxalate (Fig. 7.2, Table 7.2): results in the rat small intestine are less consistent (Chadwick et al. 1975; Saunders et al. 1975; Caspary 1977).

Although bile salts and fatty acids reproducibly increase colonic (if not jejunal) absorption of luminal oxalate in acute experiments (Table 7.2), whether this mechanism by itself explains the hyperoxaluria seen in patients after, for example, terminal ileal resection (Fig. 7.2), is not yet clear. Thus, while high doses of chenodeoxycholic acid increased ^{14}C-labelled oxalate absorption and urinary oxalate in gallstone patients (Caspary et al. 1977) no change in oxalate excretion could be detected in subjects treated either with conventional therapeutic doses of the bile acid (Caspary et al. 1977; Rampton DS 1980 unpublished work) (Fig. 7.4) or with enough ricinoleic acid to induce severe

Table 7.2. Evidence that bile salts and fatty acids increase colonic mucosal permeability to oxalate

Region of gut	Species	Technique	Agent increasing oxalate absorption	Reference
Jejunum Colon	Rat	Perfusion	– Linoleic acid	Saunders et al. 1975
Colon	Rat	Perfusion	Deoxycholate, ricinoleic and oleic acids	Dobbins and Binder 1976
Jejunum Colon	Rat	Everted sac	Glycocholate, deoxycholate, taurocholate	Caspary 1977
Colon	Rat	Everted sac	Ricinoleic acid	Schwarz et al. 1980
Colon	Rat	Short-circuit	Deoxycholate, taurocholate, ricinoleic acid	Kathpalia et al. 1984
Jejunum Colon	Rat	Perfusion	– Glycodeoxycholate	Chadwick et al. 1975
Jejunum Colon	Monkey	Perfusion	– Glycodeoxycholate	Chadwick et el. 1975
Colon	Man	Perfusion	Chenodeoxycholate	Fairclough et el. 1977

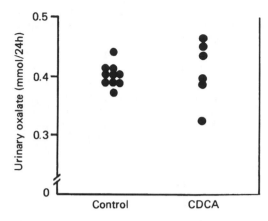

Fig. 7.4. Twenty-four hour urinary oxalate excretion in normal subjects and in patients treated for gallstones with chenodeoxycholic acid (CDCA, 15 mg/kg/day) while on a high oxalate intake. (Rampton DS, 1980 unpublished work).

watery diarrhoea (Earnest 1977). Indeed, it seems likely on theoretical grounds that, if bile salt- or fatty acid-induced increases in colonic permeability to oxalate are to cause hyperoxaluria, then soluble oxalate must be present in the lumen of the gut in raised concentration, for example as a result of coincident steatorrhoea (Fig. 7.2).

Bacterial Degradation of Oxalate

Recent work indicates that the faeces of most normal subjects contains an anaerobic bacterial species, *Oxalobacter formigenes*, capable of degrading oxalate in vitro (Allison et al. 1986). This organism could be isolated less frequently from patients with Crohn's disease and other causes of steatorrhoea (Goldkind et al. 1986). Furthermore, in these patients, as in those with a jejuno-ileal bypass (Allison et al. 1986), the rate of degradation of oxalate by faeces in vitro was reduced, possibly as a result of suppression of *Oxalobacter formigenes* by deoxycholic acid (Argenzio et al. 1985). The quantitative importance of these interesting observations in relation to the pathogenesis of EHO remains to be determined.

Site of Absorption of Oxalate

The colon appears to be the principal site of absorption of oxalate in patients with EHO. While the permeability theory obviously points to this conclusion, the pKa for fatty acids (Schwartz et al. 1980) makes hyperabsorption of oxalate consequent upon its solubilisation most likely to take place in the large bowel, where prolonged exposure to a luminal pH exceeding 6.5, and thus to an environment conducive to formation of calcium soaps, first occurs during passage of chyme down the intestine (Meldrum et al. 1972).

Experimental support for a dominant role for the colon comes from in vitro and in vivo studies in the rat and monkey showing that absorption of oxalate from the lumen of the colon is greater than that from the small intestine (Chadwick et al. 1975; Saunders et al. 1975; Caspary 1977). In clinical studies, hyperoxaluria due to hyperabsorption of dietary oxalate is rare in patients with both an ileostomy and steatorrhoea (Earnest et al. 1974; Dobbins and Binder 1977; Andersson et al. 1978; Hylander et al. 1978; Modigliani et al. 1978). In addition, in one patient with steatorrhoea, closure of his ileostomy resulted in hyperoxaluria, while in the same report, it was shown that perfusion of the colon of 3 patients with a calcium-containing solution reduced hyperoxaluria substantially (Modigliani et al. 1978). Lastly, the prolonged absorption of oxalate which follows a test meal implies its colonic rather than more proximal intestinal absorption (Barilla et al. 1978; Tiselius et al. 1981; Hofmann et al. 1983).

Mechanism of Transmucosal Oxalate Transport

Early studies suggested that intestinal absorption of oxalate occurred exclusively by passive diffusion (Binder 1974; Saunders et al. 1975; Caspary 1977; Schwartz et al. 1980), but the results may have been influenced by the choice of a calcium-free bathing medium with a consequent increase in membrane leakiness. More recent experiments using short-circuit techniques with colonic mucosa of the rat and rabbit (Freel et al. 1980; Hatch et al. 1984), and isolated brush border vesicles of the rabbit ileum (Knichelbein et al. 1986), indicate that, in these models at least, transport of oxalate is an active, energy-dependent, carrier-mediated, anion-exchange process. However, in patients with EHO, raised intraluminal concentrations of soluble oxalate together with increases in colonic mucosal permeability induced by bile salts and fatty acids are likely to make passive absorption of oxalate quantitatively more important than active transport.

Clinical Importance of Enteric Hyperoxaluria

Renal Complications

The clinical significance of EHO lies in the risks to affected patients of oxalate nephrolithiasis and of renal tubular deposition of oxalate crystals with progression to renal failure. In the event, these risks seem to be largely confined to steatorrhoeic patients with extensive small intestinal disease or resection, usually because of Crohn's disease (Dharmsathaphorn et al. 1982), or with jejuno-ileal bypass operations undertaken for treatment of morbid obesity. In the former group of patients, the incidence of oxalate-stone formation is about 6% (Deren et al. 1962; Gelzayd et al. 1968), but in the latter it may be as high as 32%, the exact figure depending in part on the length of postoperative follow-up (Dickstein and Frame 1973; O'Leary et al.

1974; Gregory et al. 1975, 1977; Stauffer 1977a). Like the overtly less-common progressive renal oxalosis (Earnest et al. 1977; Gelbart et al. 1977; Drenick et al. 1978; Das et al. 1979; Vainder and Kelly 1982), oxalate nephrolithiasis may constitute an indication for surgical restoration of intestinal continuity: both complications contribute to the decreasing popularity of intestinal bypass surgery for patients with obesity.

Oxalate nephrolithiasis in these patients, however, may not be so closely related to hyperoxaluria itself as was initially suspected (Dowling et al. 1971; Admirand et al. 1971; Smith et al. 1972). In one report, for example, amongst 87 patients with inflammatory bowel disease, urinary stones were no more common in patients with than without hyperoxaluria, and, furthermore, the stones more often contained oxalate in normoxaluric than hyperoxaluric patients (Hylander et al. 1979). In patients with Crohn's disease and with jejuno-ileal bypasses, oxalate stones are a much less common finding than hyperoxaluria (Dickstein and Frame 1973; Gregory et al. 1975, 1977; Stauffer 1977a). The same applies even more strongly to patients with EHO due to coeliac disease, pancreatic insufficiency and small bowel bacterial overgrowth, for instance, in whom there is no good evidence of an increased risk of oxalate nephrolithiasis (Dharmasathaphorn et al. 1982). Despite their tendency to hyperoxaluria, as well as hypocitraturia (Rudman et al. 1980), hypomagnesuria (Booth et al. 1963; Rudman et al. 1980), hypophosphaturia (Smith et al. 1970) and a low urinary volume, most patients with malabsorption syndromes do not develop renal stones. Elucidation of the factors which, in addition to hypocalciuria (Gregory et al. 1977; Rudman et al. 1980), militate against calculus formation in these patients (Smith 1980) would have useful therapeutic implications for the minority of individuals with intestinal disease who do acquire stones, as well of course as for patients with nephrolithiasis but no overt gut abnormalities.

Screening for Steatorrhoea with the Oxalate Loading Test

The tedium and unpleasantness of collecting faeces to measure excretion of fat in patients with suspected malabsorption, together with awareness of the direct correlation between urinary oxalate output and faecal fat excretion (Fig. 7.3; and see section on Solubility Theory, p. 106), led to the proposal that measurement of 24-h urinary excretion of oxalate after an oral oxalate load might be used as a screening test for steatorrhoea.

In the original test of Andersson and Gillberg (1977), the oxalate load was given as spinach, but because the availability of oxalate from this source varies (Earnest et al. 1975c; Hodgkinson 1977; Brinkley et al. 1981; Laker 1983), the vegetable was replaced by capsules of sodium oxalate in two further assessments of the reliability of the procedure (Rampton et al. 1979, 1984). Unfortunately, the promise of these three early studies, particularly in relation to their ability to detect, by normal urinary output of oxalate after the oral oxalate load, those patients having normal faecal fat excretion (Fig. 7.5), was not borne out in a limited further appraisal (Foster et al. 1984). Whether the oxalate-loading test will prove a genuine advance in the search for a convenient screening procedure for steatorrhoea awaits further evaluation using simpler oxalate assay techniques than were formerly available. (See Chapter 2).

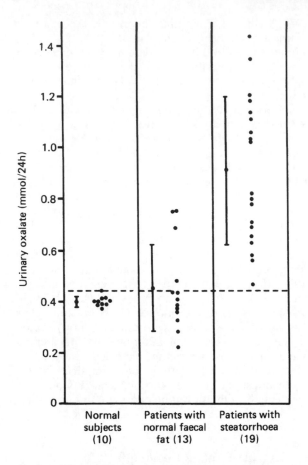

Fig. 7.5. Twenty-four hour urinary oxalate excretion in normal subjects and in patients without and with steatorrhoea on a high oxalate intake. The dashed line indicates the upper limit of oxalate excretion in normal subjects. Note that all patients with steatorrhoea had hyperoxaluria. Means ± 1 SD are shown. (Data drawn from Rampton et al. 1979).

Treatment of Enteric Hypoxaluria

Although no definite guidelines can yet be given about the indications for specific treatment of patients with EHO, it seems reasonable, in view of the rarity of nephrolithiasis in most hyperoxaluric patients, to restrict therapeutic efforts to individuals who either have established oxalate stones or have had a jejuno-ileal bypass or extensive ileal resection (see p. 110). In the remainder, treatment of the underlying intestinal disorder, for example with a gluten-free diet for patients with coeliac disease (McDonald et al. 1977) or pancreatic enzyme supplements for those with pancreatic insufficiency (Earnest 1977), is probably all that is needed. The principles of treatment for patients in whom

Table 7.3. Therapeutic possibilities in patients with enteric hyperoxaluria (EHO)

Treatment	Rationale
1. *Diet*	
Low oxalate	Reduce dietary oxalate intake
Low fat	Reduce dietary oxalate absorption
2. *Drugs*	
Divalent cations – Ca^{++}	Bind oxalate in gut lumen
Mg^{++}	Bind oxalate in gut lumen; reduce urine lithogenicity
Cholestyramine	Bind oxalate, bile salts and fatty acids in gut lumen
Aluminium hydroxide	Bind oxalate and bile salts in gut lumen
Antibiotics	Reduce bacterial oxalogenesis
Citrate	Reduce urine lithogenicity
Allopurinol	Reduce urinary oxalate and uric acid excretion
3. *Surgery*	
Ileostomy	Prevent colonic oxalate absorption
Reverse jejuno-ileal bypass	Reverse EHO

specific measures are thought necessary devolve logically from what is known about the pathogenesis of EHO (Table 7.3). It should be emphasised, however, that while the medical measures described below may reduce urinary oxalate excretion, none has been proven in longterm studies to reduce the incidence of urolithiasis in patients with EHO.

Dietary Measures (Table 7.3)

Low Oxalate Diet

The hyperabsorption of dietary oxalate which occurs in patients with EHO makes dietary oxalate restriction a rational therapy which in most patients reduces urinary output of oxalate (Chadwick et al. 1973; Stauffer et al. 1973; Earnest et al. 1974; Rampton et al. 1979; Hofmann et al. 1983). Adherence to a low-oxalate diet outside a metabolic ward, however, may be less than rigid, and it is probably more realistic simply to recommend avoidance of foods and drinks rich in oxalate (eg. spinach, beetroot, rhubarb, parsley, peanuts, chocolate, tea, coffee, cola drinks) (Hodgkinson 1977; Earnest et al. 1974; Kasidas et al. 1980; Finch et al. 1981). Furthermore, in patients with a jejuno-ileal bypass, at least, the benefits of dietary oxalate restriction may be limited by oxalogenesis from ingested protein (see p. 114).

Low-Fat Diet

Fat restriction as a treatment for EHO follows from the solubility theory of oxalate hyperabsorption. Several studies have confirmed that a low-fat diet, supplemented if necessary with medium-chain triglycerides to maintain calorie intake, reduces urinary excretion of oxalate (Andersson and Jagenburg 1974; Earnest et al. 1974; Earnest et al. 1975a,b).

Drug Therapy (Table 7.3)

Divalent Cations

It has been shown repeatedly (if not invariably (Nordenvall et al. 1983a)) that supplementation of the diet with calcium, presumably by precipitating oxalate in the gut lumen, reduces hyperoxaluria in patients with EHO (Earnest et al. 1974a,b; Caspary et al. 1977; Stauffer 1977a,b; Barilla et al. 1978; Hylander et al. 1980). Such treatment, however, requires regular monitoring of urinary calcium excretion, since in one study the beneficial effect of oral calcium on urinary excretion of oxalate was counterbalanced by a rise in urinary output of calcium so that the saturation product for urinary calcium oxalate was unchanged (Barilla et al. 1978); similar results were noted with oral magnesium supplements.

Cholestyramine

Anion-exchange resins bind oxalate (Stauffer et al. 1973; Binder 1974; Laker and Hofmann 1981), bile salts and fatty acids (Johns and Bates 1970) in vitro and would therefore be predicted to reduce urinary excretion of oxalate according to both the solubility and permeability theories. In the event, cholestyramine seems to work most consistently in patients with ileal resections and, by implication, bile-salt as well as fat malabsorption (Smith et al. 1972; Stauffer et al. 1973; Caspary et al. 1977). The decision to treat such patients with cholestyramine or other sequestrants longterm, however, should take into account the resin's detrimental effects upon steatorrhoea (Hofmann and Poley 1972) and absorption of fat-soluble vitamins and folic acid (Heaton et al. 1972; West and Lloyd 1975).

Aluminium Hydroxide

The binding properties of aluminium hydroxide in vitro resemble those of cholestyramine (Earnest 1977; Laker and Hofmann 1981), but only very limited studies of its efficacy in EHO have been published (Earnest 1977; Hofmann et al. 1981; Nordenvall et al. 1983b).

Antibiotics

Theoretically, antibiotics could either reduce urinary excretion of oxalate by inhibiting bacterial oxalogenesis from dietary protein precursors (Hofmann et al. 1983), or they could worsen hyperoxaluria by reducing colonic bacterial degradation of oxalate (Allison et al. 1986). In fact, in two studies, in which clindamycin (Nordenvall et al. 1983b) and neomycin (Hofmann et al. 1983) were given to patients with EHO, no consistent change in urinary oxalate output was observed.

Citrate

Patients with EHO are, as mentioned earlier, often hypocitraturic (Rudman et al. 1980), but there have been no systematic investigations of the effects of prolonged treatment with oral citrate on their urinary lithogenicity.

Allopurinol

As well as reducing urinary excretion of uric acid the xanthine oxidase inhibitor, allopurinol, decreases urinary output of oxalate (Scott et al. 1978; Simmonds et al. 1981; Tomlinson et al. 1985); both effects may prove beneficial in patients with recurrent calcium-oxalate stones unassociated with intestinal disease (Coe 1978; Ettinger et al. 1986; Anonymous 1987). Studies in progress indicate that allopurinol has, at most, a modest effect however on the urinary excretion of oxalate of patients with bowel disorders (D'Cruz DP, Rampton DS, Rose GA, 1988, unpublished data).

Surgical Treatment (Table 7.3)

While creation of an ileostomy in order to prevent colonic hyperabsorption of oxalate would seem to be a rather extreme treatment for EHO, recurrent oxalate nephrolithiasis and progressive renal oxalosis are definite indications for restorative surgery in patients with jejuno-ileal bypasses (see p. 111).

Summary

Enteric hyperoxaluria is by far the commonest form of secondary hyperoxaluria, occurring in patients with a variety of intestinal disorders characterised by malabsorption of fat, bile salts or both. It appears to be largely due to intestinal hyperabsorption of dietary oxalate, although in patients with a jejuno-ileal bypass oxalogenesis from dietary protein may play a role. Oxalate is hyperabsorbed as a result of two principal and complementary events: first, an increase in the luminal concentration of oxalate consequent upon precipitation of intraluminal calcium with malabsorbed fatty acids, and second, a bile-salt and fatty-acid-induced increase in colonic mucosal permeability. The contribution of reduced colonic degradation by bacteria of oxalate in malabsorptive patients is uncertain. Mucosal transport of oxalate appears to take place at least in part by an active energy-dependent process.

The clinical consequences of enteric hyperoxaluria include nephrolithiasis and, more rarely, renal tubular oxalosis with progressive renal failure: both are much less common than hyperoxaluria itself so that other factors, as yet unidentified, must play a major role in their pathogenesis. Treatment of enteric hyperoxaluria should probably be limited to patients with established nephrolithiasis or renal oxalosis, and to those with extensive ileal resections or jejuno-ileal bypasses. The therapeutic possibilities other than treatment of the underlying gut disorder include restriction of dietary oxalate and fat intake, and supplementation of such a regime with calcium, magnesium or cholestyramine. The roles of aluminium hydroxide, antibiotics, citrate and allopurinol have not yet been fully evaluated. In some patients with a jejuno-ileal bypass, surgical restoration of intestinal continuity may be necessary.

Future work should be directed to clarifying the factors determining stone formation in patients with EHO and thereby improving available therapeutic options. Some patients with idiopathic calcium-oxalate stones appear to hyperabsorb dietary oxalate (Hodgkinson 1978; Marangella et al. 1982), and the explanation of this observation, which has been made in patients without obvious gut disease, is needed. It is hoped that a continuation of the rational approach used in much of the original work reviewed here will bear fruit in the years to come.

References

Admirand WH, Earnest DL, Williams HE (1971) Hyperoxaluria and bowel disease. Trans Assoc Am Physicians 84: 307–312

Allison MJ, Cook HM, Milne DB, Gallagher S, Clayman RV (1986) Oxalate degradation by gastrointestinal bacteria from humans. J Nutr 116: 455–460

Andersson H, Gillberg R (1977) Urinary oxalate on a high oxalate diet as a clinical test of malabsorption. Lancet II: 677–679

Andersson H, Jagenburg R (1974) Fat-reduced diet in the treatment of hyperoxaluria in patients with ileopathy. Gut 15: 360–366

Andersson H, Filipsson S, Hulten L (1978) Urinary oxalate excretion related to ileocolic surgery in patients with Crohn's disease. Scand J Gastroenterol 13: 465–469

Anonymous (1987) Allopurinol for calcium oxalate stones? Lancet I: 258–259

Argenzio RA, Laicos JA, Allison MJ (1985) Role of intestinal oxalate degrading bacteria in the pathogenesis of enteric hyperoxaluria. Gastroenterology 88: 1309

Barilla DE, Notz C, Kennedy D, Pak CYC (1978) Renal oxalate excretion following oral oxalate loads in patients with ileal disease and with renal and absorptive hypercalciurias. Effect of calcium and magnesium. Am J Med 64: 579–585

Binder HJ (1973) Faecal fatty acids – mediators of diarrhoea? Gastroenterology 65: 847–850

Binder HJ (1974) Intestinal oxalate absorption. Gastroenterology 67: 441–446

Booth CC, Babouris N, Hanna S, MacIntyre I (1963) Incidence of hypomagnesaemia in intestinal malabsorption. Br Med J 2: 141–144

Briggs MH, Garcia-Webb P, Davies P (1973) Urinary oxalate and vitamin C supplements. Lancet II: 201

Brinkley L, McGuire J, Gregory J, Pak CYC (1981) Bioavailability of oxalate in foods. Urology 17: 534–538

Caspary WF (1977) Intestinal oxalate absorption. I. Absorption in vitro. Res Exp Med 171: 13–24

Caspary WF, Tönissen J, Lankisch PG (1977) 'Enteral' hyperoxaluria. Effect of cholestyramine, calcium, neomycin and bile acids on intestinal oxalate absorption in man. Acta Hepato-Gastroenterol 24: 193–200

Chadwick VS, Modha K, Dowling RH (1973) Mechanism for hyperoxaluria in patients with ileal dysfunction. New Engl J Med 289: 172–176

Chadwick VS, Elias E, Bell GD, Dowling RH (1975) The role of bile acids in the increased intestinal absorption of oxalate after ileal resection. In: Matern S, Hackenschmidt J, Back P, Gerok W (eds) Advances in Bile Acid Research. Schattauer, Stuttgart, pp 435–440

Coe FL (1978) Hyperuricosuria and calcium oxalate nephrolithiasis. Kidney Int 13: 418–426

Das S, Joseph B, Dilk AL (1979) Renal failure owing to oxalate nephrosis after jejuno-ileal bypass. J Urol 121: 506–509

Deren JJ, Porush JG, Levitt MF, Khilani MT (1962) Nephrolithiasis as a complication of ulcerative colitis and regional enteritis. Ann Intern Med 56: 843–853

Dharmsathaphorn K, Freeman D, Binder HJ, Dobbins JW (1982) Increased risk of nephrolithiasis in patients with steatorrhoea. Dig Dis Sci 27: 401–405

Dickstein SS, Frame B (1973) Urinary tract calculi after intestinal shunt operations for the treatment of obesity. Surg Gynaecol Obstet 136: 257–260

Dobbins JW, Binder HJ (1976) Effect of bile salts and fatty acids on the colonic absorption of oxalate. Gastroenterology 70: 1096–1100

Dobbins JW, Binder HJ (1977) Importance of the colon in enteric hyperoxaluria. New Engl J Med 296: 298–301

Dowling RH, Rose GA, Sutor DJ (1971) Hyperoxaluria and renal calculi in ileal disease. Lancet I: 1103–1106

Drenick EJ, Stanley TM, Border WA et al. (1978) Renal damage with intestinal bypass. Ann Intern Med 89: 594–599

Dvorackova I (1966) Tödliche Vergiftungnach intravenöser Verabreichung von Natriumoxalat. Arch Toxikol 22: 63–67

Earnest DL, Johnson G, Williams HE, Admirand WH (1974) Hyperoxaluria in patients with ileal resection: an abnormality in dietary oxalate absorption. Gastroenterology 66: 1114–1122

Earnest DL, Williams HE, Admirand WH (1975a) Treatment of enteric hyperoxaluria with calcium and medium chain triglyceride. Clin Res 23: 130

Earnest DL, Williams HE, Admirand WH (1975b) A physicochemical basis for treatment of enteric hyperoxaluria. Trans Assoc Am Physicians 88: 224–234

Earnest DL, Williams HE, Admirand WH (1975c) An explanation for the presence or absence of hyperoxaluria in patients with jejuno-ileal bypass. Gastroenterology 68: 1072

Earnest DL (1977) Perspectives on incidence, aetiology and treatment of enteric hyperoxaluria. Am J Clin Nutr 30: 72–75

Earnest DL, Campbell J, Hunt TK et al. (1977) Progressive renal failure with hyperoxaluria: a new complication of jejuno-ileal bypass. Gastroenterology 72: 1054

Earnest DL (1979) Enteric hyperoxaluria. Adv Intern Med 24: 407–427

Ettinger B, Tang A, Citron JT, Livermore B, Williams T (1986) Randomized trial of allopurinol in the prevention of calcium oxalate calculi. New Engl J Med 315: 1386–1389

Fairclough PD, Feest TG, Chadwick VS, Clark ML (1977) Effect of sodium chenodeoxycholate on oxalate absorption from the excluded human colon – a mechanism for 'enteric' hyperoxaluria. Gut 18: 240–244

Finch AM, Kasidas GP, Rose GA (1981) Urine composition in normal subjects after oral ingestion of oxalate-rich foods. Clin Sci 60: 411–418

Fitzpatrick JM, Kasidas GP, Rose GA (1981) Hyperoxaluria following glycine irrigation for transurethral prostatectomy. Br J Urol 53: 250–252

Foster PN, Will EJ, Kelleher J, Losowsky MS (1984) Oxalate loading tests to screen for steatorrhoea: an appraisal. Clin Chim Acta 144: 155–161

Frascino JA, Vanamee P, Rosen PP (1970) Renal oxalosis and azotemia after methoxyfluorane anaesthesia. New Engl J Med 283: 676–679

Freel RW, Hatch M, Earnest DL, Goldner A (1980) Oxalate transport across the isolated rat colon. A re-examination. Biochim Biophys Acta 600: 838–843

Gelbart DR, Brewer LL, Weinstein AB (1977) Oxalosis and chronic renal failure after intestinal bypass. Arch Intern Med 137: 239–243

Gelzayd EA, Breuer RI, Kirsner JB (1968) Nephrolithiasis in inflammatory bowel disease. Amer J Dig Dis 13: 1027-1034

Goldkind R, Cave DR, Jaffin B, Bliss CM, Allison MJ (1986) Bacterial oxalate metabolism in the human colon: a possible factor in enteric hyperoxaluria. Gastroenterology 90: 1431

Gregory JG, Starkloff EB, Miyai K, Schoenberg HW (1975) Urologic complications of ileal bypass operation for morbid obesity. J Urol 113: 521–524

Gregory JG, Park KY, Schoenberg HW (1977) Oxalate stone disease after intestinal resection. J Urol 117: 631–634

Hatch M, Freel RW, Goldner AM, Earnest DL (1984) Oxalate and chloride absorption by the rabbit colon: sensitivity to metabolic and anion transport inhibitors. Gut 25: 232–237

Heaton KW, Lever JV, Barnard D (1972) Osteomalacia associated with cholestyramine therapy for post-ileectomy diarrhoea. Gastroenterology 62: 642–646

Hodgkinson A (1977) Oxalic acid in biology and medicine. Academic Press, New York

Hodgkinson A (1978) Evidence of increased oxalate absorption in patients with calcium-containing stones. Clin Sci 54: 291–294

Hofmann AF, Thomas PJ, Smith LH et al. (1970) Pathogenesis of secondary hyperoxaluria in patients with ileal resection and diarrhoea. Gastroenterology 58: 960

Hofmann AF, Poley JR (1972) Role of bile acid malabsorption in pathogenesis of diarrhoea and steatorrhoea in patients with ileal resection. I. Response to cholestyramine or replacement of dietary long chain triglyceride by medium chain triglyceride. Gastroenterology 62: 918–934

Hofmann AF, Tacker M, Fromm H, Thomas PJ, Smith LH (1973) Acquired hyperoxaluria and intestinal disease. Evidence that bile acid glycine is not a precursor of oxalate. Mayo Clin Proc 48: 35–42

Hofmann AF, Schmuck G, Scopinaro N et al. (1981) Hyperoxaluria associated with intestinal bypass surgery for morbid obesity: occurrence, pathogenesis and approaches to treatment. Int J Obesity 5: 513–518

Hofmann AK, Laker MF, Dharmsathaphorn K, Sherr HP, Lorenzo D (1983) Complex pathogenesis of hyperoxaluria after jejunoileal bypass surgery. Oxalogenic substances in diet contribute to urinary oxalate. Gastroenterology 84: 293–300

Hylander E, Jarnum S, Juel Jensen H, Thale M (1978) Enteric hyperoxaluria; dependence on small intestinal resection, colectomy and steatorrhoea in chronic inflammatory bowel disease. Scand J Gastroenterol 13: 577–588

Hylander E, Jarnum S, Frandsen I (1979) Urolithiasis and hyperoxaluria in chronic inflammatory bowel disease. Scand J Gastroenterol 14: 475–479

Hylander E, Jarnum S, Neilsen K (1980) Calcium treatment of enteric hyperoxaluria after jejunoileal bypass for morbid obesity. Scand J Gastroenterol 15: 349–352

Jeghers H, Murphy R (1945) Practical aspects of oxalate metabolism. New Engl J Med 233: 208–215

Johns WH, Bates TR (1970) Quantification of the binding tendencies of cholestyramine. II: mechanism of interaction with bile salt and fatty acid salt anions. J Pharm Sci 59: 329–333

Kasidas GP, Rose GA (1980) Oxalate content of some common foods: determination by an enzymatic method. J Hum Nutr 34: 255–266

Kathpalia SC, Favus MJ, Coe FL (1984) Evidence for size and charge perm-selectivity of rat ascending colon: effects of ricinoleate and bile salts on oxalic acid and neutral sugar transport. J Clin Invest 74: 805–811

Knichelbein RG, Aronson PS, Dobbins JW (1986) Oxalate transport by anion exchange across rabbit ileal brush border. J Clin Invest 77: 170–175

Laker MF, Hofmann AF (1981) Effective therapy of enteric hyperoxaluria: in vitro binding of oxalate by anion exchange resins and aluminium hydroxide. J Pharm Sci 70: 1065–1067

Laker MF (1983) The clinical chemistry of oxalate metabolism. Adv Clin Chem 23: 259–297

Marangella M, Fruttero B, Bruno M, Linari F (1982) Hyperoxaluria in idiopathic stone disease: further evidence of intestinal hyperabsorption of oxalate. Clin Sci 63: 381–385

Mekhjian HS, Phillips SF (1970) Perfusion of the canine colon with unconjugated bile acids: the effect on water and electrolyte transport, morphology and bile acid absorption. Gastroenterology 59: 120–129

Mekhjian HS, Phillips SF, Hofmann AF (1971) Colonic secretion of water and electrolytes induced by bile acids: perfusion studies in man. J Clin Invest 50: 1569–1577

Meldrum SJ, Watson BW, Riddle HC, Brown RL, Sladen GE (1972) pH profile of gut as measured by radiotelemetry capsule. Br Med J 2: 104–106

McDonald GB, Earnest DL, Admirand WH (1977) Hyperoxaluria correlates with fat malabsorption in patients with sprue. Gut 18: 561–566

Modigliani R, Labayle D, Aymes C, Denvil R (1978) Evidence for excessive absorption of oxalate by the colon in enteric hyperoxaluria. Scand J Gastroenterol 13: 187–192

Nordenvall B, Backman L, Larsson L, Tiselius H-G (1983a) Effects of calcium, aluminium, magnesium and cholestyramine on hyperoxaluria in patients with jejunoileal bypass. Acta Chir Scand 149: 93–98

Nordenvall KJ, Hallberg D, Larsson L, Nord CE (1983b) The effect of clindamycin on the intestinal flora in patients with enteric hyperoxaluria. Scand J Gastroenterol 18: 177–181

O'Leary JP, Thomas WC, Woodward ER (1974) Urinary tract stone after small bowel bypass for morbid obesity. Am J Surg 127: 142–147

Parry MF, Wallach R (1974) Ethylene glycol poisoning. Am J Med 57: 143–150

Rampton DS, Kasidas GP, Rose GA, Sarner M (1979) Oxalate loading test: a screening test for steatorrhoea. Gut 20: 1089–1094

Rampton DS, Breuer NF, Vaja SG, Sladen GE, Dowling RH (1981) Role of prostaglandins in bile salt-induced changes in rat colonic structure and function. Clin Sci 61: 641–648

Rampton DS, McCullough DA, Sabbat JS, Salisbury JR, Flynn FV, Sarner M (1984) Screening for steatorrhoea with an oxalate loading test. Br Med J 288: 1419, 1728

Ribaya JD, Gershoff SN (1982) Factors affecting endogenous oxalate synthesis and its excretion in feces and urine of rats. J Nutr 112: 2161–2169

Rofe AM, Conyers RAJ, Bais R, Edwards JB (1979) Oxalate excretion in rats injected with xylitol or glycollate: stimulation by phenobarbitone pre-treatment. Aus J Exp Biol Med Sci 57: 171–176

Rudman D, Dedonis JL, Fountain MT et al. (1980) Hypocitraturia in patients with gastrointestinal malabsorption. New Engl J Med 303: 657–661

Saunders DR, Sillery J, McDonald GB (1975) Regional differences in oxalate absorption by rat intestine: evidence for excessive absorption by the colon in steatorrhoea. Gut 16: 543–554

Schmidt KH, Hagmaier V, Hornig DH, Vuilleumier JP, Rutishauser G (1981) Urinary oxalate excretion after large intakes of ascorbic acid in man. Am J Clin Nutr 34: 305–311

Schwartz SE, Stauffer JQ, Burgess LW, Cheney M (1980) Oxalate uptake by everted sacs of rat colon. Biochim Biophys Acta 596: 404–413

Scott R, Paterson PJ, Mathieson A, Smith M (1978) The effect of allopurinol on urinary oxalate excretion. Br J Urol 50: 455–458

Simmonds NA, Van Acker KJ, Potter CF, Webster DR, Kasidas GP, Rose GA (1981) Influence of purine content of diet and allopurinol on uric acid and oxalate excretion levels. In: Smith LH, Robertson WG, Finlayson B (eds) Urolithiasis: clinical and basic research. Plenum, New York, pp 363–378

Smith LH, Hofmann AF, McCall JT, Thomas PJ (1970) Secondary hyperoxaluria in patients with ileal resection and oxalate nephrolithiasis. Clin Res 18: 541

Smith LH, Fromm H, Hofmann AF (1972) Acquired hyperoxaluria, nephrolithiasis and intestinal disease. New Engl J Med 286: 1371–1375

Smith LH (1980) Enteric hyperoxaluria and other hyperoxaluric states. In: Coe FL (guest ed), Brenner BM, Stein JH (eds) Contemporary Issues in Nephrology vol 5. Churchill Livingstone, New York, pp 215–238

Stauffer JQ, Humphreys MH, Weir JG (1973) Acquired hyperoxaluria with regional enteritis after ileal resection. Ann Intern Med 79: 383–391

Stauffer JQ (1977a) Hyperoxaluria and calcium oxalate nephrolithiasis after jejunoileal bypass. Am J Clin Nutr 30: 64–71

Stauffer JQ (1977b) Hyperoxaluria and intestinal disease. The role of steatorrhoea and dietary calcium in regulating intestinal oxalate absorption. Dig Dis Sci 22: 921–928

Thomas DW, Edwards JB, Gilligan JE, Lawrence JR, Edwards RG (1972) Complications following intravenous administration of solutions containing xylitol. Med J Aust 1: 1238–1246

Tomlinson B, Cohen SL, Al-Khader A, Kasidas GP, Rose GA (1985) Further reduction of oxalate by allopurinol in stone formers on low purine diet. In: Schwille PO, Smith LH, Robertson WG, Vahlensieck W (eds) Urolithiasis and related clinical research. Plenum, New York, pp 513–516

Tiselius HG, Ahlstrand C, Lundström B, Nilsson MA (1981) (^{14}C) oxalate absorption by normal persons, calcium oxalate stone formers, and patients with surgically disturbed intestinal function. Clin Chem 27: 1682–1685

Vainder M, Kelly J (1982) Renal tubular function secondary to jejuno-ileal bypass. J Am Med Assoc 235: 1257–1258

West RJ, Lloyd JK (1975) The effect of cholestyramine on intestinal absorption. Gut 16: 93–98

Zarembski PM, Hodgkinson A (1969) Some factors influencing the urinary excretion of oxalic acid in man. Clin Chim Acta 25: 1–10

Chapter 8

Mild Metabolic Hyperoxaluria. A New Syndrome

G. A. Rose

Introduction

The term "mild metabolic hyperoxaluria" (MMH) is being used to define a group of patients with the following characteristics:

1. The patients all have hyperoxaluria compared to healthy individuals of the same age and sex. Usually the urinary oxalate is only slightly elevated although in one or two cases very high levels have been seen
2. The word "metabolic" signifies that urinary glycollate is also elevated, approximately matching the urinary excretion of oxalate. This raised urinary glycollate proves that the hyperoxaluria is not due to over-ingestion or over-absorption of dietary oxalate (Rose and Kasidas 1979)
3. The clinical course is of severe and recurrent urolithiasis with stones made of pure calcium oxalate
4. In some cases it has been possible to measure oxalate in plasma and urine samples taken simultaneously and to show that both levels are raised. Hence the hyperoxaluria is due to raised plasma level and not to primary renal loss of oxalate. Plasma glycollate has not been measured, but it seems likely that it too is raised
5. Some patients respond dramatically to only 10 mg per day of pyridoxine. Others require higher doses and yet others will not respond even to 800 mg per day. This and other factors are described below

Historical

Six years ago we described two rather odd cases of recurrent urolithiasis and hyperoxaluria (Harrison, Kasidas and Rose 1981). One was a young man aged

23 years and the other was a young woman of 21 years of age. The latter also had renal tubular acidosis for which she had been treated with alkalis. The cause in both cases was thought to be metabolic rather than a diet high in oxalate because urinary glycollate values were also elevated. Although the degree of hyperoxaluria was much less than is usual in primary hyperoxaluria, because the latter may respond to high dose pyridoxine (Gibbs and Watts 1970), high doses of this particular vitamin were given to both patients. Urinary oxalate and glycollate values fell to normal and remained in the normal range long after the pyridoxine was stopped, quite unlike what occurs in primary hyperoxaluria. Moreover the female actually showed a response to the quite low dose of 10 mg/day.

These two cases alerted us to the possibility that low doses of pyridoxine might have been successful in both of these cases. Since then we have seen 11 other cases of MMH, some of which have been reported elsewhere (Kasidas and Rose 1984; Gill and Rose 1986). In these cases treatment has always been started with low doses of pyridoxine increasing up to 800 mg/day/70 kg body weight until a response is obtained. However, the responses have been remarkably variable. Here it is proposed to review the condition as it appears today.

Clinical Details of Patients

Brief clinical details of the 13 cases are shown in Table 8.1. It can be seen that there are 9 males and 4 females, a sex ratio that is similar to that seen in primary hyperoxaluria (Rose 1988b). The degree of hyperoxaluria is usually much less than in the primary condition but, nevertheless, the stone-recurrence rate has been very rapid until urinary oxalate has been controlled. Urinary glycollate is often slightly greater than urinary oxalate, unlike the findings in primary hyperoxaluria.

Unlike primary hyperoxaluria, although MMH can commence in childhood, age of onset is apparently often in adult life. However, some caution is required in making this statement. While it is certainly true that onset of urolithiasis may be in adult life, urinary oxalate is not measured prior to onset of symptoms of urolithiasis and one cannot therefore be certain of the duration of the hyperoxaluria. However, it seems likely that the hyperoxaluria is not present from infancy because urolithiasis may strike with remarkably severity in adult life.

Link with Other Conditions

There has been in this series a high incidence of other conditions which by themselves can cause calcigerous stones. Remarkably, one patient also had primary hyperparathyroidism due to a single parathyroid adenoma, one also

Table 8.1. Clinical details of patients

Pretreatment levels 24-h Urinary oxalate (mmol)	24-h Urinary glycollate (mmol)	Name	Sex	Age at first stone symptoms	Age when hyperoxaluria recognised	Highest dose of pyridoxine given (mg/day)	Response to pyridoxine	Other medical conditions	Years of follow-up	Other treatment
0.61, 0.69	0.80, 0.84	NY	F	21	22	100	Complete. Did not relapse on stopping	Renal tubular acidosis	10	Sodium bicarbonate
0.78, 0.83	0.79	DM	M	23	37	400	Complete. Did not relapse on stopping		2.5	
0.74, 0.60	0.44, 0.47	GS	M	24	34	10	Complete. Relapsed on stopping. Returned again on restarting	Hypertension	6	Hypertension drugs
0.39 → 1.68	0.98	AD	M	19	25	800	Complete at first. Relapsed on stopping. Responded again but gradually became resistant		6	
0.58	0.59	RJ	M	26	51	20	Complete but relapsed on stopping. Responded completely a second time	Primary hyperparathyroidism	3	Neck exploration and removal of parathyroid adenoma
0.63	0.75	SM	M	18	18	800	Temporary response and then relapsed. Responded on stopping high dose	Medullary sponge kidney Hypercalciuria	3	Bendrofluazide
0.48	0.65	AM	M	34	35	20	Response still being assessed	Idiopathic hypercalciuria	3	
a0.49, 0.60	0.43, 0.46	RS	F	$2\frac{10}{12}$	$3\frac{5}{12}$	100	Responded to 10 mg/day but later needed 100 mg/day		2.5	
a0.88, 0.81	0.42, 0.31	ME-A	F	$1\frac{7}{12}$	$1\frac{10}{12}$	300	Partial response only			
0.69, 0.67	0.86, 0.60	SS	M	30	57	800	Partial response to 50 mg/day but relapsed. Dose raised to 800 mg/day with no further response	Idiopathic hypercalciuria	3	Bendrofluazide
b0.96, 0.68	0.74, 0.56	JJ	F	5	12	20	Complete: very short follow-up		5/12	
b1.5→3.5	2.3→54.5	BD	M	9	10	400	Complete response to 10 mg/day followed by partial relapse even on high doses		4	
0.43	0.39	RL	M	57	63	50	Complete			

a per 2.5 mmol creatinine
b per 10 mmol creatinine

had medullary sponge kidney, one also had renal tubular acidosis (RTA) and two had idiopathic hypercalciuria. An important point arises from these observations: in patients with calcium-oxalate urolithiasis there is an absolute indication to measure urinary oxalate even if a diagnosis other than hyperoxaluria is apparent. Thus, in the case of RJ (Gill and Rose 1986), who also had primary hyperparathyroidism, it is likely that the cause of his stones was the hyperoxaluria and not the hyperparathyroidism since the stones finally proved to be almost pure calcium oxalate, an event that is extremely unlikely in primary hyperparathyroidism (Rose 1988a). In the case of NY (Harrison et al. 1981) who also had RTA, calcium oxalate was only 27% of the stone and hence one may go even further and say that the presence of only a small proportion of calcium oxalate in a stone may be an indication to measure urinary oxalate. It is now clear that over-absorption of dietary oxalate may accompany over-absorption of dietary calcium. However, it is equally clear that one cannot assume that hyperoxaluria is dietary in origin when idiopathic hypercalciuria is present and it is important in these cases to measure urinary glycollate to find out whether or not MMH is present.

The link between hyperoxaluria and RTA is of considerable interest. Attention has been drawn elsewhere (Kasidas and Rose 1984) to the occurrence of RTA in a group of patients, under the care of Professor O. Wrong, who actually passed stones made of calcium oxalate and proved to have MMH. Of the 7 patients, all of whom were adults, 5 were males. They were treated with moderate doses of pyridoxine in the range of 50–300 mg/day and each responded with a fall to normal in urinary oxalate and glycollate. In some cases the RTA was familial and the MMH also proved to be familial suggesting that the aetiology of the latter was different from the other cases of MMH described here.

The high incidence of other causes of stones in patients with MMH is certainly remarkable and one is bound to wonder whether, say, medullary sponge kidney, idiopathic hypercalciuria or primary hyperparathyroidism can in some way induce the development of MMH. This seems rather unlikely because when the primary hyperparathyroidism was cured surgically this did not lead to remission of MMH and neither did treatment of hypercalciuria with thiazide diuretics.

Response to Pyridoxine

The responses to pyridoxine described below must raise the question of patient compliance. Did the patients take the doses of vitamin prescribed? To check on this point pyridoxic acid has been measured in the patients' urine samples and it has been gratifying to find invariably that the expected excretions have been found.

The variation in response to pyridoxine has been quite remarkable. At one extreme there is GS (Kasidas and Rose 1984) who responded to only 10 mg/day of pyridoxine, relapsed on stopping and responded a second time with the same dose. His urinary oxalate and glycollate have remained normal on this dose for 6 years. RJ (Gill and Rose 1986) responded twice to only 20 mg/day and relapsed in the interval when the dose was stopped. Thereafter the low dose of pyridoxine was continued and his urinary oxalate and glycollate have remained normal for 2 years. JJ also showed an excellent response to 20 mg/day of pyridoxine, but she returned to the USA and is lost to follow-up. RL was given 50 mg/day and his urinary levels of oxalate and glycollate fell to normal and have remained there for 3 years.

Other patients have been much more resistant to pyridoxine. Thus AD (Gill and Rose 1986) has been followed up for 6 years so far (Fig. 8.1). At first he responded to only 20–50 mg/day of pyridoxine, urinary oxalate and glycollate falling right into the normal range. When the dose was stopped he quickly relapsed but responded to the same dose a second time. However, he became increasingly resistant to pyridoxine as urinary levels of oxalate and glycollate persistently remained slightly above normal despite a stepwise rise in dose of pyridoxine over 6 years to 800 mg/day, the highest dose that can be used with safety. Urinary pyridoxic acid showed that he always took the prescribed dose of pyridoxine and it seems he has developed an amazing drug resistance. In August 1987 the pyridoxine was stopped to investigate the possibility that the treatment was no longer effective. A month later urinary oxalate was up to

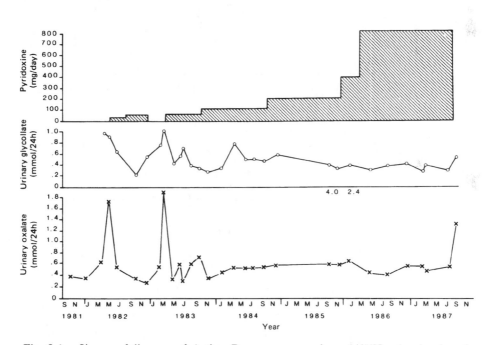

Fig. 8.1. Six-year follow-up of Arthur D, a young man born 14/1/57, who developed mild metabolic hyperoxaluria while under observation, but must almost certainly have had it previously. He responded twice to low doses of pyridoxine, but quickly relapsed when the dose was stopped in November 1982. After March 1983 urinary oxalate and glycollate levels remained slightly raised despite increasing doses of pyridoxine. On stopping high doses of pyridoxine the hyperoxaluria immediately increased further. Follow-up continues.

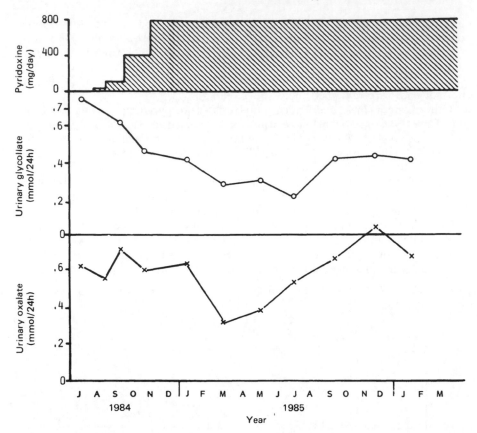

Fig. 8.2. Two-year follow-up of Simon M, a young man born 7/5/65 who had both medullary sponge kidney and mild metabolic hyperoxaluria. The latter failed to respond to low doses of pyridoxine, but did respond to 800mg/day after some months of treatment, only to relapse a few months later. Compliance was confirmed by measuring urinary pyridoxic acid. After stopping the pyridoxine, urinary oxalate and glycollate values fell to normal. The latest oxalate value (July 1987) was 0.34 mmol/24 h.

1.02 mmol/24 h and glycollate was up to 0.52 mmol/24 h. Treatment with pyridoxine was therefore restarted.

SM (Gill and Rose 1986) was also remarkably resistant to pyridoxine as there was no response until the dose was raised to 800 mg/day (Fig. 8.2). After about 4 months of this high dose the urinary oxalate did fall to normal only to be followed by steady relapse during the next 9 months. Eventually the pyridoxine was discontinued completely in October 1986 and then, surprisingly, the urinary oxalate and glycollate levels promptly fell to normal and remained there up to July 1987 and follow-up continues.

The responses to pyridoxine are detailed in Table 8.1 and may be summarised thus:

1. Four subjects made a good response to low doses of pyridoxine and did not relapse

2. Two subjects made good responses to high doses and did not relapse when the pyridoxine was stopped

3. Four subjects made good responses to low doses of pyridoxine but subsequently needed higher doses

4. One subject made temporary response to high doses of pyridoxine and then relapsed, and yet remitted when the dose was stopped

5. One subject has made little or no response to increasing doses

This remarkable diversity of responses seems to suggest two points. First, there is some similarity to the response to high doses of pyridoxine seen in primary hyperoxaluria. It has been reported (Rose 1988b) that about one-third of cases of primary hyperoxaluria show complete response, about one-third show no response and about one-third show partial or temporary response. Second, MMH seems to start and stop in certain patients for reasons which are not clear, but will be considered more fully below.

Aetiology

There seems little doubt that MMH can begin at almost any age (see Table 8.1) and that it can remit. Thus DM and NY (Harrison et al. 1981) remitted for long periods of time after high doses of pyridoxine and it seems that SM has remitted after stopping 800 mg/day. BD (Fig. 8.3) became worse during a period of observation prior to treatment while several other cases (see above) required increasing doses to control hyperoxaluria. The case of AD (Gill and Rose 1986) is particularly instructive. When first measured the urinary oxalate was normal, but it subsequently rose to high levels. No-one reading the history of recurrent urolithiasis from pure calcium-oxalate stones could doubt that he must have had hyperoxaluria at an earlier period. It therefore seems clear from considering these patients that the condition of MMH begins and stops spontaneously. Presumably there is an environmental factor which causes this to happen but what it is is not at all clear. We have considered the contraceptive pill, intake of wine containing antifreeze, occupation and food intake but have been unable to establish any connection between these items and MMH. The possibility that patients start and stop taking pyridoxine without telling the investigator has also been considered, but excluded by the finding that urinary pyridoxic-acid measurements have always indicated total compliance (see above).

Pyridoxine deficiency or dependency can cause clinical conditions other than hyperoxaluria. These include sideroblastic anaemia (Horrigan and Harris 1964), epilepsy (Hunt et al. 1954), mental retardation with homocystinuria (Barber and Spaeth 1964) or cystathioninuria (Frimpter et al. 1963). None of the patients with MMH have shown any signs of these conditions. In some cases it has been possible to measure red-cell-transaminase activities. These have been normal prior to treatment and have shown the normal degree of

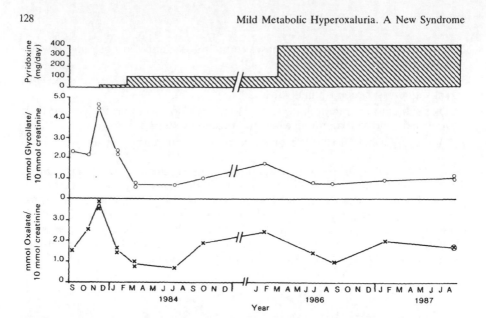

Fig. 8.3. Four-year follow-up of Bryn D, born 3/3/73, a boy under the care of Professor Barratt (see Chapter 6) with mild metabolic hyperoxaluria of increasing severity until partial response to only 10 mg/day of pyridoxine. He relapsed, but showed further partial response to 400 mg/day of pyridoxine. Compliance was verified by urinary pyridoxic-acid measurements.

increase on adding pyridoxal phosphate. It is therefore difficult to postulate a general disturbance of pyridoxine metabolism. However, pyridoxal phosphate is the coenzyme for a considerable number of different enzyme systems and presumably only the system handling glyoxylate is affected.

One must be bound to wonder if MMH is a forme fruste of primary hyperoxaluria. In favour of this suggestion are:

1. In both conditions response to pyridoxine may be complete, partial or zero
2. In both conditions males predominate over females

However, firmly against the suggestion are:

1. MMH may apparently commence at any age and may remit, both of these situations being unknown in primary hyperoxaluria
2. No two cases have been seen so far in a family and this seems to make a purely genetic cause rather unlikely. However, it does not exclude the possibility of an environmental factor acting on genetically predisposed individuals

Have Others Seen MMH?

Murthy et al. (1982) treated with 10 mg/day of pyridoxine for 180 days 12 recurrent stone formers who had mild metabolic hyperoxaluria. They found reductions in 24-h urinary-oxalate levels. However, they did not measure glycollate levels and it is therefore impossible to be sure that they were seeing metabolic hyperoxaluria. Balcke et al. (1983) treated 12 idiopathic stone formers with 300 mg/day of pyridoxine for 6 weeks. Twenty-four-hour urinary oxalate fell in every case, mean values pre and post-treatment being 0.38 and 0.336 mmol respectively. However, with their normal range of 0.228–0.412 mmol/24 h only 7 of the pretreatment values were above normal. Although mean 24-h urinary glycollate fell from 208 to 153 mmol, none was above the normal range before treatment. Again, this condition does not seem to be the same as MMH although it could be a variant. Jaeger et al. (1986) gave 300 mg/day of pyridoxine for 8 weeks to 10 idiopathic stone formers with mild hyperoxaluria of unknown origin and found that the urinary oxalate fell substantially in 8 subjects, and very slightly in another, the mean level falling from 0.509–0.382 mmol/24 h. Again, glycollate values were not reported. They have, however, been measured since and are said to have been raised (personal communication) so that these patients probably had MMH.

Effect of Pyridoxine on Stone Formation

There seems little doubt that this group of patients has made fewer stones since attending here than previously. In particular NY, DM, GS, AD, RJ, RS, BD, and RL have made no more stones since starting pyridoxine as compared with severe recurrent stone-growth in some of these cases before starting pyridoxine. While it is tempting to attribute these successes to pyridoxine therapy and a consequent fall in urinary oxalate, and this may be a correct interpretation, one should bear in mind the non-specific improvement arising from attendance at a metabolic-stone clinic (Hosking et al. 1983). However, it is of interest that SM, who continued to have hyperoxaluria for most of the follow-up period despite pyridoxine therapy, did suffer further stone growth, and it therefore seems reasonable to suppose that the fall in urinary oxalate induced by pyridoxine is beneficial in reducing further stone formation.

References

Balcke P, Schmidt P, Zazgornik J, Kopsa H, Minar E (1983) Pyridoxine therapy in patients with renal calcium oxalate calculi. Proc Eur Dial Transplant Assoc Eur Ren Assoc 20: 417–421
Barber GW, Spaeth GL (1967) Pyridoxine therapy in homocystinuria. Lancet I: 337
Frimpter GW, Haymovitz A, Horwith M (1963) Cystathioninuria. N Engl J Med 268: 333–339

Gibbs DA, Watts RWE (1970) The action of pyridoxine in primary hyperoxaluria. Clin Sci 38: 277–286

Gill HS, Rose GA (1986) Mild metabolic hyperoxaluria and its response to pyridoxine. Urol Int 41: 393–396

Harrison AR, Kasidas GP, Rose GA (1981) Hyperoxaluria and recurrent stone formation apparently cured by short courses of pyridoxine. Br Med J 282: 2097–2098

Horrigan DL, Harris JW (1964) Pyridoxine responsive anaemia. Adv Intern Med 121: 163–174

Hosking DH, Erickson SB, Van Den Berg CJ (1983) The stone clinic effect in patients with idiopathic calcium urolithiasis. J Urol 130: 1115–1118

Hunt AD, Stokes SJ, McCrory WW, Stroud HH (1954) Pyridoxine dependency: a report of a case of intractable convulsions in an infant. Paediatrics 13: 140–145

Jaeger H, Portmann L, Jacquet A-F, Burckhardt P (1986) La pyridoxine peut normaliser l'oxalurie dans la lithiasè rénale idiopathique. Schweiz Med Wochenschr 116: 1783–1786

Kasidas GP, Rose GA (1984) Metabolic hyperoxaluria and its response to pyridoxine. In: Ryall R, Brockis JG, Marshall V, Finlayson B (eds) Urinary stone. Churchill-Livingstone, Melbourne, pp 138–147

Murthy MSR, Farooqui S, Talwar HS, Thind SK, Nath R, Rajendran L, Bapna BC (1982) Effect of pyridoxine supplementation on recurrent stone formers. Int J Clin Pharm Therap Toxicol 20: 434–437

Rose GA (1988a) Primary hyperparathyroidism. In: Wickham JEAW, Buck AC (eds) Renal tract stone: metabolic basis and clinical practice. Churchill-Livingstone, London, In press

Rose GA (1988b) The role of pyridoxine in stone prevention of hyperoxaluric patients. Urology Annual (In press)

Rose GA, Kasidas GP (1979) New enzymatic method for measurement of plasma and urinary glycollate and its diagnostic value. In: Gasser G, Vahlensieck W (eds) Pathogenese und Klinik der Harnsteine VII. Steinkopf-Verlag, Darmstadt, pp 252–260

Chapter 9

Oxalate Crystalluria

P. C. Hallson

Introduction

Crystalline deposits found in urine have long been associated with the formation of renal stones. By the 1930s most of the common crystalline deposits in urine had been identified and the acidic or alkaline conditions in which they are usually seen established (Ball and Evans 1932). Phosphates of lime and magnesium were known to be soluble in acid urine and deposited in alkaline urine. Similarly calcium hydrogen phosphate was seen in faintly acid or alkaline conditions, whilst calcium oxalate in the form of dumb-bell or envelope-shaped crystals appeared in acid and alkaline urine.

Whilst there is still little to dispute in these early findings, methods of investigating urinary crystals have improved markedly since that time, driven by the persuasive belief that crystalluria and stone formation are somehow related. Among the first advances was the recognition that urine samples for crystal examination must be maintained at body temperature if spurious precipitation of urinary sediments due to cooling is to be avoided (Dyer and Nordin 1967). Urine samples left to stand may change in pH (usually towards alkalinity), which may give rise to further error. These observations enabled an investigation of crystalluria in circumstances closer to physiological conditions.

Dyer and Nordin's observations were followed by what may have been the first attempt at precise quantitative measurement of crystalluria in fresh urine at 37°C. The method was described by Robertson and utilised the Coulter electronic particle counter to measure numbers, sizes and volumes of calcium crystals present in urine (Robertson 1969). The usual examination of urinary sediments, until this time, was confined to light microscopy after centrifugation of the urine, which was frequently a 24-h specimen left both to cool and to stand.

Crystal Types seen by Light Microscopy at 37°C

The microscopic appearances of both calcium oxalate and calcium phosphate crystals are outlined below. As will be apparent later, a study of calcium phosphate crystals is pertinent to enquiries into calcium oxalate crystalluria.

Calcium Oxalate

Calcium oxalate crystals are found in urines of normal and stone-forming individuals usually in the form of easily identifiable octahedral crystals often referred to as "envelope" oxalate crystals on account of their appearance. These envelopes vary in size and although those usually seen lie between 2 and 10 μm, crystals as large as 100 μm have been found. Another less common form of calcium oxalate gives the appearance of dumb-bells (or hour-glasses or ovals). These crystals are themselves polycrystalline agglomerates (Berg et al. 1976). Although the envelope oxalate crystal is commonly thought of as the dihydrate and the dumb-bell forms as the monohydrate there is evidence that dihydrate crystals may also appear in dumb-bell or hour-glass form (Berg et al. 1976). Dumb-bell crystals have a similar size distribution to the envelope forms.

Both types of calcium oxalate crystals are able to cluster into rigid aggregates. Some investigators have referred to these as "microstones" (Von Sengbusch and Timmerman 1957) and they may well be the precursors of calcium oxalate stones, frequently resembling small stones of this material.

Aggregates may arise by coalescence between crystal faces possibly aided by heterogeneous substances or homogeneous bridging. Agglomeration of individual crystals is opposed by an electrical (Zeta) potential on or near the crystal surface which serves to maintain particles in suspension. Some reputed inhibitors of aggregation are believed to function by increasing this potential, hence increasing the repulsive forces between particles (Scurr and Robertson 1986). Crystals which succeed in approaching sufficiently closely might adhere. Sticking may be aided by viscous forces due to surface-bound layers of macromolecules or by a recrystallisation process which yields a bridge between adjacent crystal faces (Rushton et al. 1981; Finlayson 1978).

Calcium Phosphate

Amorphous forms of calcium phosphate are very commonly seen in urines above pH 6.5. A variety of complex phosphates have been characterised, some being basic, others acidic (Hallson 1977). In urine important ones appear to be octocalcium phosphate ($Ca_4H (PO_4)_3$) and hydroxyapatite ($Ca_5(OH)(PO_4)_3$) (Robertson et al. 1968). In contrast to calcium oxalate all are soluble in acetic acid. Brushite is an acidic calcium phosphate ($CaHPO_4$) sometimes seen as "stellar" crystals which appear nearly always between pH 6 and 7. Calcium oxalate (or the oxalate anion) has been found coprecipitated with amorphous calcium phosphate and this topic is discussed later.

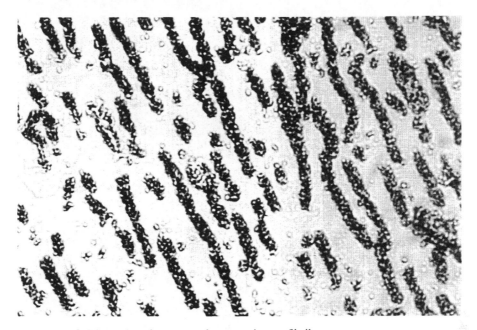

Fig. 9.1. Calcium phosphate crystals appearing as fibrils.

Speculation as to the involvement of large molecules in the aggregation of calcium oxalate crystals already mentioned extends also to calcium phosphate crystals. Microscopic examination of centrifuged calcium phosphate crystals, particularly from high osmolality urines, shows that these crystals are often stuck together in clusters. After ultrafiltration (pore size 12 000 daltons) the clustering disappears yielding unattached crystals which readily disperse across a microscope slide. The effect is attributable to the high molecular weight Tamm–Horsfall mucoprotein, normally present in urine, which polymerises into an insoluble form at higher urine osmolalities (Hallson and Rose 1979). If uromucoid is re-added to ultrafiltered urine before precipitation the clustering of calcium phosphate crystals is restored. Also if, during microscopic examination of a calcium phosphate precipitate, the coverslip is slid horizontally across the slide and entrapped calcium phosphate deposit, clumped calcium phosphate crystals are often seen to resolve into fibrils aligned in the direction that the coverslip was pulled (see Fig. 9.1; Hallson and Rose unpublished observations 1987). It is possible therefore that crystal nucleation, adhesion or both occur upon filamentous strands of the insoluble polymerised uromucoid, and account for what is seen. Whilst this effect is most noticeable in centrifuged deposits, similar fibrils with amorphous calcium phosphate attached may also be seen in uncentrifuged urine sediments. Aggregated envelope calcium oxalate crystals are also occasionally seen as linear or chain-like structures (Fig. 9.2). Whether this is connected with the above observations relating to calcium phosphate is unclear but there is evidence for the role of uromucoid in triggering calcium oxalate and calcium phosphate crystal nucleation (Hallson and Rose 1979) and also agglomeration (Scurr and Robertson 1986).

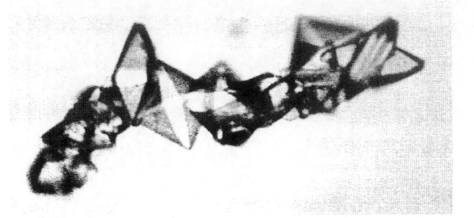

Fig. 9.2. Linear or chain-like aggregate of envelope oxalate crystals.

Investigating Crystalluria

Studies of crystalluria in freshly voided urine collected in Dewar (Thermos) flasks to maintain the sample at body temperature began at St Peter's Hospitals, London in 1974. Samples came mainly from patients attending the out-patient stone clinic at these hospitals, but specimens from normal individuals were also examined (Hallson and Rose 1976).

After urine had been collected the contents of the flasks were mixed by inversion and a 25-ml sample removed into a universal container with a conical base. The sample was centrifuged in a warm room (maintained near 37°C) and examined for crystals using a microscope fitted with a 37°C-heated stage. These studies yielded much information relating to the incidence and morphology of crystals in fresh warm urine (Hallson and Rose 1976, 1977).

From these initial studies two disadvantages became clear, firstly that a high proportion of urines seemed crystal-free and secondly, that no quantitative estimate of crystal volume was possible.

Inducing Crystallisation

A means whereby naturally occurring urinary crystals could be obtained from crystal-free urines has several potential uses both from clinical and experimental viewpoints. Crystals of calcium oxalate may be formed by straightforward slow infusion of calcium and oxalate ions into urine. Unfortunately, crystals generated in this way are not only dissimilar in themselves but different again to those found in natural urine (Hallson and Rose 1978). Crystals of cubic form appeared when calcium chloride and sodium oxalate were infused into water, whilst dilute urine yielded hexagonal, dendritic and amorphous calcium oxalate. The natural envelope crystals were rarely formed. The absence of naturally occurring crystals in dilute urine is consistent with the observation that quite small changes in the composition of crystallising media can produce striking alterations in crystal form or habit (Mullen 1972). Other studies of urinary crystal formation support this observation (Welshman and McGeown 1972).

Generating Crystals by Evaporation

One tried and proven method for inducing crystalluria is to evaporate crystal-free fresh urine at 37°C using a rotary-film evaporator with a suitable high-vacuum pump (Hallson and Rose 1978). This procedure has a number of points in its favour.

1. The urine can be maintained at body temperature throughout
2. The process is analogous to the abstraction of water in the collecting tubules of the kidneys. Relative concentrations of urinary solutes therefore remain constant. This is not the case in methods where calcium and oxalate are added to urine
3. The final concentration (or osmolality) may be decided in advance. This enables a comparison of crystalluria in different urine samples at a constant osmolar concentration
4. A considerable advantage of the evaporation technique is that crystals so formed are microscopically indistinguishable from their counterparts seen in untreated urines. For example, envelope calcium oxalate crystals, whether single or aggregated, are identical to those seen in untreated urines. This applies to all other crystal types commonly encountered such as cystine, uric acid, stellar calcium phosphate, magnesium ammonium phosphate, etc.
5. Evaporation enables crystals to be found in a majority of urines where otherwise they would be seen only in a minority, of the order of 10%–15% of untreated urines (Hallson and Rose 1976, 1978).

In practice a known weight of urine of measured osmolality is rapidly evaporated at 37°C until the weight falls to a value corresponding to the (predecided) final osmolality. This may be, for example 800, 1000 or 1200 mOsm/kg. A portion of the evaporated urine (or an aliquot in the case of quantitative crystal measurements) is now centrifuged at 37°C and examined by microscopy as described.

A recent innovation has been the application of reverse osmosis hollow-fibre membrane to concentrate calcium and oxalate to supersaturation and early precipitation levels (Azoury et al. 1986). Whilst this technique has not yet been applied to urine, it does provide a close simulation of collecting-tubule water resorption in a flowing system.

Quantitative Crystal Measurements

The problem of quantifying crystalluria has been attempted in several ways, some of which give relative values of crystalluria whilst others yield absolute amounts.

Determining Relative Amounts of Calcium Oxalate Crystals

The simplest means of measuring relative amounts of calcium oxalate crystals is probably by microscopy and crystal counting. For example, the numbers of envelope oxalate crystals may be counted over a number of fields after the crystal deposit (derived from a known volume of urine) is suspended in a fixed volume of mother liquor, homogenised and dispersed across the microscope slide. The method is aided by use of a graticule eye-piece. A good correlation was found using the above procedure with relative volumes of calcium oxalate crystals determined by ^{14}C-labelled oxalic acid isotope counting (Hallson et al. 1982). This isotopic technique is certainly more precise. A trace amount of isotopically labelled oxalate (usually ^{14}C-labelled oxalic acid) is introduced into the urine prior to crystal formation. On crystallisation a proportion of the isotope-labelled compound will be incorporated into the calcium oxalate precipitate. The crystals may be separated by centrifugation, washed (saturated ammonium oxalate) and the radioactivity (counts/time) measured. Only relative changes in crystal formation are possible unless the endogenous oxalate content of the urine prior to precipitation has been determined. In the latter case the concentration (in μmol/l) of oxalate crystals may be calculated, provided that the counts for a given volume of urine, prior to crystallisation, have also been measured.

The handling of radioactive materials is however a necessary feature of this procedure.

Absolute Quantities of Calcium Oxalate Crystals

A well-tried technique to assess sizes, numbers and volumes of urinary calcium crystals was devised by W. G. Robertson (Robertson 1969) and uses the Coulter particle counter. In outline the method is as follows. Urine is collected into a warm Dewar vessel, and all later operations are carried out at 37°C. The urine is prefiltered using a 74 μm sieve to remove large particles which may block the orifice through which urine flows during counting. After sieving, the sample is divided into two aliquots (of 45 ml) one of which (A) is treated with ethylenediaminetetra-acetic acid (EDTA) solution (5 ml) buffered at the pH of the urine and incubated in a shaking water-bath at 37°C for 20 min. To the second aliquot (B) is added sodium-chloride solution (5 ml) to keep this solution iso-osmotic with the first. Sample B is then incubated for 20 min at 37°C. The EDTA solution dissolves all calcium crystals present in sample A; in sample B calcium crystals are retained. Aliquot A, therefore, contains only urinary debris (epithelial and red cells, casts etc) or non-calcium crystals whilst aliquot B contains all crystals plus debris. The Coulter counter was used to determine the crystal numbers in a series of size ranges from 3.8 μm to 48.6 μm. Subtracting counts obtained for aliquot A from those for aliquot B in each size-range yielded the numbers of crystals in each size-range. In addition total volumes of crystals per size-range can be ascertained from the number and size measurements.

This application of the Coulter counter is ingenious but has some severe limitations.

1. The technique does not differentiate calcium oxalate crystals from those of calcium phosphate. Both forms may be found across the physiological pH range (see Figs. 9.19a,b) although calcium phosphate is rarely present below pH 5.6. Even microscopy of urinary sediments in combination with the Coulter counter would be inadequate for this purpose since amorphous forms of calcium phosphate and oxalate are indistinguishable by light microscopy

2. A more serious problem is the inability to count particles below 3.8 μm diameter. Certainly most amorphous calcium phosphate and amorphous calcium oxalate particles fall below this size range and even some envelope oxalate crystals do so (see Fig. 9.3). These, evidently, would not be counted. On several occasions a mass of envelope calcium oxalate crystals, just resolvable by light microscopy have appeared in urinary sediments, a high proportion of which were less than the 3.8 μm diameter threshold (Hallson and Rose, unpublished data). Hence a large volume of very small crystals is likely to be underestimated

3. The binding or clumping observed in calcium phosphate crystals mentioned earlier constitutes a further problem since it may well give rise to errors in crystal sizing

4. A practical difficulty with the Coulter counter technique is that initial filtration with a 74 μm pore sieve does not restrict fairly large numbers of blood and epithelial cells present in some urines. These clump during incubation and obstruct the counting orifice giving spurious results

Despite these limitations the Coulter counter remains a useful instrument for the investigation of calcium crystalluria although the limitations must always be remembered.

Chemical Analysis of Urinary Crystals

A need therefore remained to determine the extent of calcium oxalate crystalluria across the entire range of crystal sizes. The very low level of calcium oxalate present in most urine samples had, until recently, defied chemical methods of analysis.

In 1985 a method was described for continuous-flow assay of urinary oxalate using immobilised oxalate oxidase derived from barley seeds. The method operated on the AutoAnalyser I (Kasidas and Rose 1985). The oxidase enzyme used, which yields hydrogen peroxide with oxalate, is severely inhibited by the salt concentration in the urine range. This necessitated a fiftyfold dilution of the urine together with the use of a chromogenic system with a high extinction coefficient. These problems having been overcome, the system measures urinary oxalate in the range of mmol/l and oxalate as μmol/l in low-salt aqueous solution (see Chaps. 2, 4).

Shortly after its application to urinary oxalate the method was applied successfully to the measurement of calcium oxalate crystals in urinary sediments (Hallson and Rose 1988). Calcium oxalate crystal solutions are almost salt-free so dilution is unnecessary.

In outline the procedure is as below. All operations are carried out at 37°C. A known aliquot of urine (which may be either untreated or post-evaporation) is placed in a conical-based universal container, centrifuged (3000 rpm for 5 min) and the supernatant decanted. The sediment is washed with saturated calcium sulphate, centrifuged and the wash liquid removed. The washing is repeated and the wash liquid removed again. Hydrochloric acid (0.05 M, 1 ml) is added to the crystal-containing sediment and after agitation to break up the precipitate, the tube is allowed to stand for 1 h to ensure complete dissolution of calcium crystals. The sample is now centrifuged to remove insoluble debris. An aliquot of 0.5 ml of the supernatant is buffered to pH 3.5, the operating pH of the assay, by addition of 4.5 ml citrate buffer, and centrifuged. The supernatant is assayed for oxalate using the AutoAnalyser procedure as described. A portion of the crystal solution in hydrochloric acid is retained for analysis for phosphate by a standard centrifugal analyser procedure. The latter gives a measure of calcium phosphate crystal concentration.

The precision of the crystal concentration determinations was assessed by duplicate crystal estimations on evaporated urines. When duplicate samples were derived from a single urine evaporation the average difference from the mean of duplicates was 4.51% for oxalate and 4.10% for phosphate. When samples were evaporated separately the average difference between the mean of duplicate values for calcium oxalate was 10.69%, and for calcium phosphate 8.19%. The evaporation procedure, as may be expected, reduced the level of precision to some extent.

Recovery of calcium oxalate crystals was determined as follows. Samples of 5, 10, 15 and 20 ml were withdrawn from a single urine after evaporation. The highest volume sample (20 ml and found to contain 0.164 μmol of calcium oxalate) was set at 100% recovery. The corresponding recoveries for the 5, 10 and 15 ml aliquots were 104.9%, 101.2% and 97.6% respectively.

In practice, urines for crystal analysis should be sampled as quickly as possible as crystal growth rates of calcium oxalate and phosphate, especially in concentrated and post-evaporation urines, can introduce substantial error (Hallson and Rose 1988; Rose and Sulaiman 1984). The range of concentrations of calcium oxalate crystals seen in studies of crystalluria using the assay method described above, varies from less than 1 μmol/l to 500 μmol/l, the lower values being the most frequent (Fig. 9.20a,b). The mean value is close to 5 μmol/l. These figures are consistent with earlier measurements of calcium oxalate crystalluria in which urine crystals were recovered on hydrophobic filters (Hallson 1977). Here the observed range in 13 samples was from 0 to 420 μmol/l with a mean of 41 μmol/l. This procedure was cumbersome and required large urine volumes. Incomplete removal of residual urine might partly account for the higher mean crystal concentration of calcium oxalate found.

Recent Studies of Crystalluria

Nearly all the data in the following pages have been derived from studies of crystalluria begun at St Peter's Hospitals in London in 1984 using light microscopy, urine evaporation and the AutoAnalyser assay of calcium oxalate crystals as described. The studies cover both morphological and quantitative aspects of oxalate crystalluria following investigation of 1200 urines from stone-formers and normal individuals. For convenience the bipyramidal form of calcium oxalate dihydrate crystals are referred to as "envelope oxalate" crystals, whilst the dumb-bell, oval and hour-glass forms are collectively referred to as "dumb-bell" crystals.

A Wescor 5100c Vapour pressure osmometer was used to determine urine osmolalities. Urines subjected to evaporation were concentrated to 1200 mOsm/kg ($\pm$ 25 mOsm/kg). pH was measured by glass electrode.

Crystal Sizes

Figure 9.3a,b shows the distribution of the maximum size observed of unaggregated envelope oxalate crystals seen in 59 untreated and 212 evaporated urine samples from stone-forming subjects. Sizes were estimated with the aid of a calibrated graticule eye-piece. The most frequent sizes were from 5 to 10 μm, consistent with earlier findings (Robertson 1969).

A feature of Fig. 9.3a,b is the relative absence of envelope oxalate crystals above 15 μm in post-evaporation urines compared with untreated ones. A probable explanation is that crystals in evaporated urine were separated from the mother liquor immediately after formation. On the other hand, crystals found in untreated urines had an unknown residence time in the bladder prior to voiding in which crystal growth may well have taken place. Rapid crystal growth in calcium oxalate crystals has already been shown (Hallson and Rose 1988) and, excepting instances when many new small crystals form, this is probably accompanied by increasing crystal size.

"Eroded" Envelope Oxalate Crystals

The time which urine spends in the bladder may also cause morphological changes. The envelope appearance of calcium oxalate dihydrate crystals is sharply defined and easily recognisable. In some instances, however, the crystals are seen with poorly defined edges giving the appearance of erosion. This raises the possibility that eroded crystals may have formed in urine under conditions initially favourable to the growth of this crystal habit, but were not voided until a change in the conditions of bladder contents had taken place. If this change was a rise in urinary pH to above 6.0 (where envelopes are

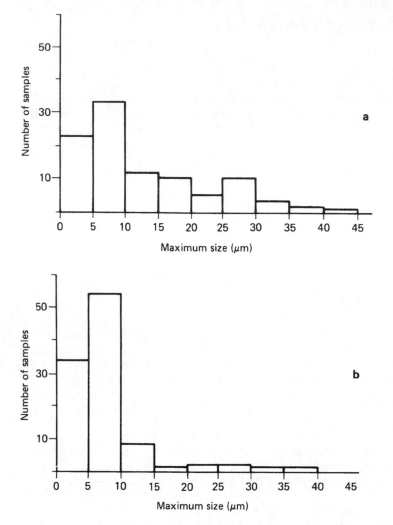

Fig. 9.3. **a** The size distribution of envelope oxalate crystals in untreated urines. The values shown are those of the largest-sized crystal seen in the urine sample. **b** The size distribution of envelope oxalate crystals in evaporated urines. The values shown are those of the largest-sized crystal seen in the urine sample.

much less abundant) it might be that partial dissolution of the envelopes occurs, giving them an eroded appearance. In fact in the 7 cases where such erosion has been noted in untreated urines, all urine pH values were over 6.0 (mean 6.75). This lends support to the possibility that some dissolution of crystals had taken place.

The foregoing findings concerning size and erosion of envelope oxalate crystals illustrate a very important aspect of urinary crystal studies. The point is that crystals seen after voiding are by no means necessarily present, even in embryonic form, when the urine is secreted into the bladder. Furthermore, crystals which have developed in the bladder urine may themselves alter before voiding. Urine crystals, thus, only reflect the potential of a urine sample to generate certain types of crystal. This, in turn, may relate to the capacity of the urine to promote stone growth.

A point in favour of the evaporation method for inducing crystalluria from crystal-free urine is that crystals so formed may be examined immediately after formation. This simulates urine in the renal pelvis immediately after the removal of water in the collecting tubules. On the other hand, bladder urine has been subject to various changes in crystal structure.

The Importance of Urinary Oxalate in Calcium Oxalate Crystalluria

In a solution with as many components as urine and where numerous interactions and complex compounds are possible involving ions, small and macro molecules and particulate matter, forecasting the effects upon crystalluria of changes in urine composition is particularly difficult. This has not, however, deterred attempts to do so (Robertson et al. 1968; Werness et al. 1985). Calculations of this type have indicated that small increases in urinary oxalate concentration are much more likely to promote crystalluria than substantial increases in urine calcium (Robertson and Nordin 1976; Robertson et al. 1981).

These predictions have been supported by experimental evidence. Addition of buffered calcium chloride to aliquots of urine produced calcium oxalate crystals only at abnormally high calcium levels. On the other hand, increasing the oxalate concentration to the top of the normal range caused precipitation (Robertson and Nordin 1969). When the volumes of calcium oxalate crystals in stone-former urines were determined using the Coulter counter technique, these were found to be closely related to the urine oxalate concentration (Robertson and Peacock 1980). On the other hand, no relationship was apparent between crystalluria and the urinary concentration of calcium. To account in part for these findings the existence of a *soluble*, non-ionised calcium oxalate complex has been suggested (Robertson and Nordin 1969; Robertson and Peacock 1980). However, as will be clearer later, the influence of the urinary concentration of calcium cannot be entirely neglected.

In the recent studies at St Peter's Hospitals the urinary concentration of calcium oxalate crystals was compared with the corresponding urinary level of oxalate in urines from stone-forming and normal subjects, using both untreated urines and post-evaporation samples (Fig. 9.4). A gradual increase in concentration of oxalate crystal takes place up to about 0.40 mmol/l which is close to the upper limit of the normal range (approximately 0.07–0.30 mmol/l). Here there is a point of inflexion after which the concentration of calcium oxalate crystal rapidly increases with the urinary concentration of oxalate. This agrees with the earlier findings (Robertson and Nordin 1969). Furthermore, the severity of stone formation as reflected by the stone episode rate per patient, also shows a very similar relationship to the urinary concentration of oxalate crystal, again with a steep rise appearing near the top of the normal oxalate range (Robertson et al. 1979).

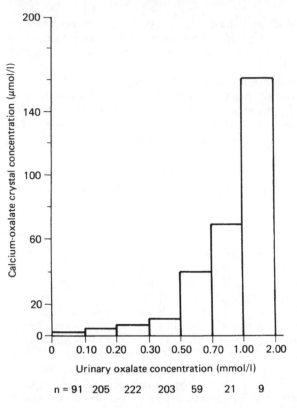

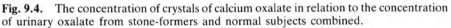

Fig. 9.4. The concentration of crystals of calcium oxalate in relation to the concentration of urinary oxalate from stone-formers and normal subjects combined.

Within the normal range, therefore, increases in urinary concentrations of oxalate are likely to be far less deleterious than those occurring above this range. At this point it should be noted that in data such as those shown in Fig. 9.4, derived from a large number of urine samples, some correlation is likely between oxalate concentration in urine and the concentrations of many ionic and molecular species which may either promote or inhibit crystallisation, the most clear of which is calcium. Nevertheless there is convincing reason to believe that this dramatic increase in calcium oxalate crystalluria with urinary oxalate concentration is primarily attributable to the latter. Evidence for this will be apparent later.

Microscopy of Urinary Oxalate Crystals

Figure 9.5 shows how often envelope oxalate crystals are seen in normal and stone-former urines with various urinary oxalate concentrations. This figure takes no account of crystal concentration and shows no point of inflexion as

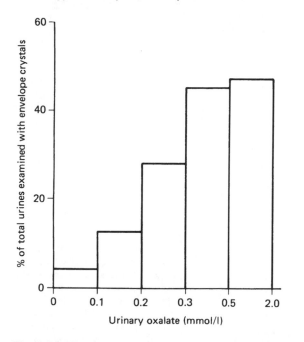

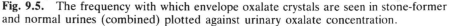

Fig. 9.5. The frequency with which envelope oxalate crystals are seen in stone-former and normal urines (combined) plotted against urinary oxalate concentration.

does Fig. 9.4. Whilst this histogram indicates a direct relation between urinary oxalate and envelope oxalate crystalluria, it is noticeable that only about one half of the numbers of urines examined showed this type of crystal at the highest concentration of urinary oxalate. This may well be due, at least in part, to calcium oxalate crystals manifesting in forms other than the envelope variety especially at higher urinary pH levels, where an amorphous form predominates.

Urinary calcium concentration correlates to some extent with urinary oxalate concentration and hence its contribution to the crystalluria shown in Fig. 9.5 is unknown. If, however, urine samples are selected from a low and limited urinary calcium concentration range then the effect of calcium in promoting crystalluria should be held constant. When Fig. 9.5 is redrawn to include only those samples with a urinary calcium concentration from 0 to 4 mmol/l, the direct correlation of envelope oxalate sightings and urinary oxalate remains (Fig. 9.6) and does show inflexion at 0.3 mmol/l.

When the frequency with which envelope calcium oxalate crystals are found in a restricted, but high, urinary calcium range, is considered a different picture emerges (Fig. 9.7). Although there is an absence of crystals at the lowest urinary oxalate range, in the remaining bands the frequency is uniformly fairly high. This tentatively suggests that calcium cannot be ignored as a factor in oxalate crystal formation and this matter is re-examined later.

Envelope oxalate crystals are the form most frequently seen in normal and stone-forming urine, the dumb-bell type being a much rarer sighting. Whilst dumb-bell oxalate crystals are frequently found in the urine of hyperoxaluric

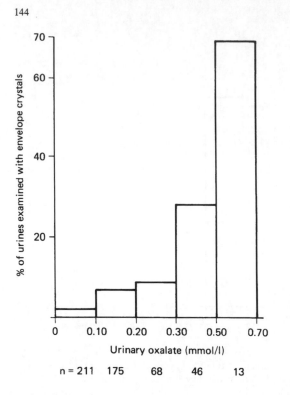

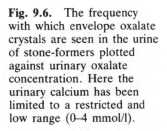

Fig. 9.6. The frequency with which envelope oxalate crystals are seen in the urine of stone-formers plotted against urinary oxalate concentration. Here the urinary calcium has been limited to a restricted and low range (0–4 mmol/l).

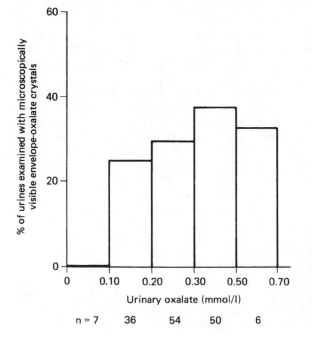

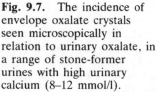

Fig. 9.7. The incidence of envelope oxalate crystals seen microscopically in relation to urinary oxalate, in a range of stone-former urines with high urinary calcium (8–12 mmol/l).

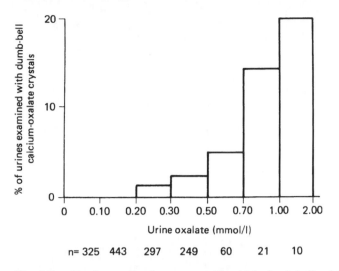

Fig. 9.8. The increasing frequency with which dumb-bell calcium oxalate crystals are seen as the oxalate concentration rises in stone-former urines. Note the absence of dumb-bells at low levels of oxalate in urine.

subjects it is our experience, and that of others, that this crystal habit is not confined to metabolic hyperoxalurics but may have dietary origins also (Catalina and Cifuentes 1970; Berg et el. 1976). Dumb-bell crystals are seen occasionally in the urine of normal and non-hyperoxaluric stone-formers. Like envelope oxalate crystals, the frequency with which dumb-bell oxalate crystals are found rises with urinary oxalate concentration (Fig. 9.8). One important difference between envelope and dumb-bell crystals is that the latter rarely appear at low concentrations of oxalate in urine. It is possible that their formation is related to the calcium/oxalate ratio of the mother liquor. Dumb-bell crystals may be seen exclusively or admixed with the envelope form. They also have the capacity to aggregate into rigid clusters, sometimes in combination with envelope type crystals.

Aggregation of Envelope Oxalate Crystals

In view of the foregoing findings linking the concentrations of urinary oxalate and calcium oxalate crystals, and the frequency of both envelope and dumb-bell crystals of calcium oxalate it may be little surprise that the frequency with which aggregated envelope crystals of oxalate are seen under the microscope is also proportional to the concentrations of urine oxalate (Fig. 9.9). The rise is steep and aggregates are seen in about one-third of urines with an oxalate concentration above 1.0 mmol/l.

Little difference was present in the morphology of such aggregates from normal subjects and stone-formers (data pertaining to size of aggregate and the number of crystals comprising it were based upon the largest aggergate

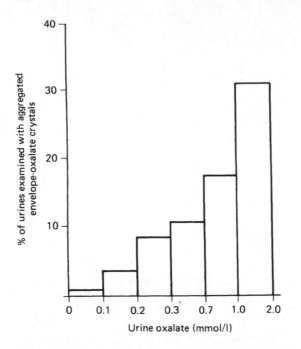

Fig. 9.9. The incidence of aggregates of envelope oxalate crystals plotted against the concentration of urine oxalate in stone-formers and normal individuals.

seen in each urine sample). The mean sizes were similar in the normal and stone-forming groups (79 μm and 65 μm respectively), as also was the mean number of crystals in the aggregate (16 and 15 respectively). No correlation was found between the size of the largest aggregates seen and the urine oxalate concentration of the corresponding urines. The mean pH of urine samples containing aggregates was 5.9 for normal subjects and 6.1 for stone-formers neither being significantly different from pH 6.4 which was the mean for all samples in the present study.

Whilst 1.2% of urines from normal subjects contained aggregated envelope crystals and 3.3% of non-hyperoxaluric stone-formers did so, this figure rose to 33% for patients with metabolic hyperoxaluria. This again emphasises the influence of urinary oxalate concentration upon calcium oxalate crystallisation and aggregate formation.

Role of Urinary Calcium Concentration in Oxalate Crystalluria

That urinary calcium plays a part in promoting calcium oxalate crystals has already been alluded to in connection with Fig. 9.7. Figure 9.10 reinforces this evidence showing a relationship between the frequency of observation of

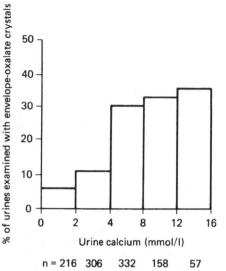

Fig. 9.10. The frequency of appearance of envelope oxalate crystals in relation to the concentration of calcium in the urine.

envelope oxalate crystals and urinary calcium concentration. The representation in Fig. 9.10 is valid, providing no correlation exists between urinary calcium and oxalate in the urine samples which comprise the raw data. Since some relationship is likely, further examinations have been carried out (Figs. 9.11,

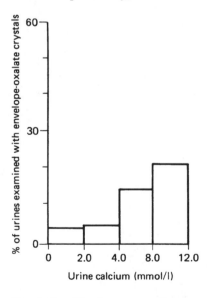

Fig. 9.11. The frequency of appearance of envelope oxalate crystals in relation to calcium concentration in the urine in stone-formers. Here all urines with an oxalate level over 0.2 mmol/l were excluded.

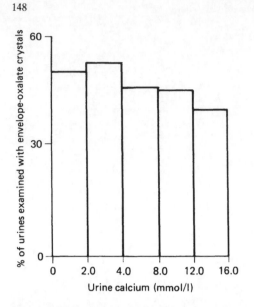

Fig. 9.12. Incidence of envelope oxalate crystals plotted against the concentration of calcium in the urine in stone-formers. In this histogram the oxalate level of all samples of urine lay between 0.5 and 2.0 mmol/l.

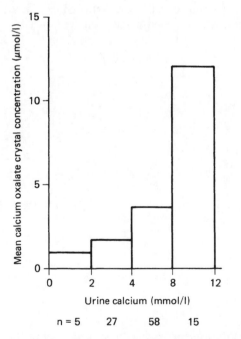

Fig. 9.13. Concentration of calcium oxalate crystals in relation to calcium concentration in the urine in stone-formers in a restricted urinary oxalate range (0.1–0.2 mmol/l).

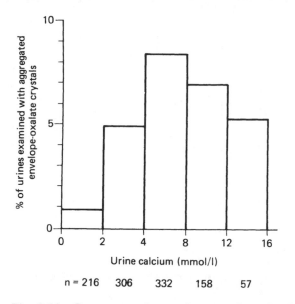

Fig. 9.14. Percentage of stone-former urines showing envelope oxalate aggregates plotted against urinary calcium concentration.

9.13) in which the oxalate concentration has been restricted to urines at the lower end of the urinary concentration range. At this level, changes in calcium oxalate crystalluria with urine oxalate are minimal (see Fig. 9.4) and, furthermore, any correlation between calcium and oxalate is also likely to be minimal. Despite this limitation, both the rate of appearance of envelope oxalate crystals (Fig. 9.11) and the urinary concentration of calcium oxalate crystals (Fig. 9.13) rise with urinary calcium.

Thus calcium cannot be neglected as a factor in crystalluria. However, comparison of Fig. 9.13 with Fig. 9.4 shows that the range of urinary calcium values found was less effective in promoting oxalate crystalluria compared with that of the corresponding range of urinary oxalate.

At the highest range of urinary oxalate (0.5–2.0 mmol/l) no increase in the incidence of envelope oxalate crystals arises with increasing urinary calcium (Fig. 9.12). Here the overwhelming effect of high urinary oxalate concentration results in a high incidence of envelope oxalate crystals even at the lowest band of urinary calcium concentration. Although Fig. 9.12 shows crystals of calcium oxalate in only about 50% of samples this implies that envelope oxalate crystals have been seen in a very high proportion of urines in the pH range favouring their formation (ie, pH 4.8–6.2).

We have already seen (Fig. 9.9) a clear relationship between incidence with which envelope oxalate aggregates are seen and their corresponding concentrations of urinary oxalate. A similar histogram (Fig. 9.14) shows the relationship of urinary calcium to such aggregates. At lower calcium concentrations there is a suggestion that increases in the concentration of urinary calcium might encourage aggregation but the reverse is apparently true at the higher calcium concentrations.

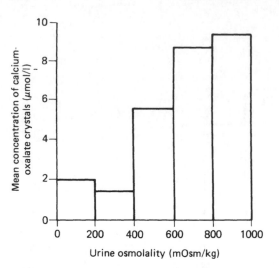

Fig. 9.15. Concentration of calcium oxalate crystals in relation to osmolality in stone-former urines.

Urine Osmolality and Calcium Oxalate Crystalluria

As urine osmolality rises the volume of precipitated calcium oxalate increases (Fig. 9.15). The increasing crystalluria results from the increases in all ionic, molecular and particle concentrations especially those of calcium and oxalate, where the probability of exceeding the [Ca] [Ox] solubility product also rises. Note, however, that in the highest osmolality range, mean concentration of calcium oxalate crystals is modest (approximately 9 μmol/l) by comparison with that induced by the highest urinary oxalate concentration (160 μmol/l; Fig. 9.4). This again shows the potency of oxalic acid in generating oxalate crystalluria. On the other hand, there may be no marked correlation between osmolality and urinary oxalate.

Figure 9.16 shows the frequency with which envelope oxalate crystals appear at different osmolality ranges. This figure may be compared with Fig. 9.5. It is again apparent that increases in urinary oxalate are more likely than increases in overall concentration to yield envelope oxalate crystals.

Osmolality and Calcium Phosphate Crystalluria

In contrast to concentrations of calcium oxalate crystals, no clear relationship could be discerned between concentrations of calcium phosphate crystals and osmolality. Figure 9.17 is a histogram of the frequency with which calcium phosphate crystals are seen against the osmolality ranges in which the crystals

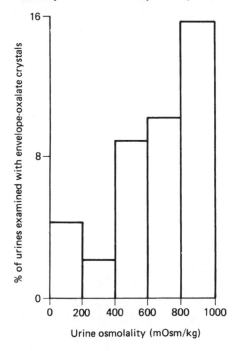

Fig. 9.16. The percentage of urines from stone-formers which show envelope oxalate crystals plotted against urine osmolality.

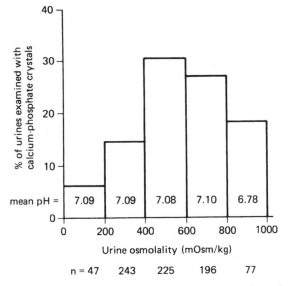

Fig. 9.17. The percentage of stone-former urines showing calcium phosphate crystals plotted against urine osmolality.

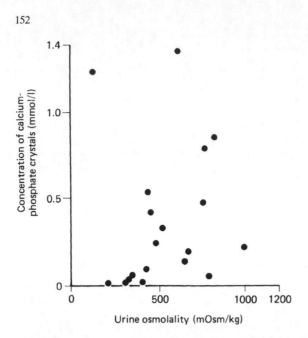

Fig. 9.18. The concentrations of calcium phosphate crystals in 19 stone-former urines with the pH range 7.0–7.2 plotted against urine osmolality.

arose. Being independent of the widely varying concentrations of calcium phosphate crystals, this figure is a more sensitive means of detecting the effects of osmolality upon the formation of calcium phosphate crystals.

Urine pH is a strong determinant of calcium phosphate precipitation and nearly certainly accounts for the fall in phosphate crystals in the highest osmolality class (800–1000) where the mean pH is significantly lower (see Fig.

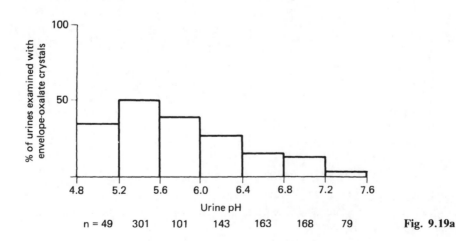

Fig. 9.19a

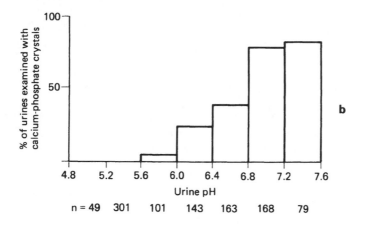

Fig. 9.19. **a** The percentage of stone-former urines showing envelope oxalate crystals plotted against the pH of the urine. **b** The percentage of stone-former urines showing calcium phosphate crystals plotted against the pH of the urine.

9.17). Since pH exerts a strong influence in this way it may be possible to observe an osmolality effect upon phosphate crystalluria within a restricted pH range. Figure 9.18 shows the distribution of concentration of calcium phosphate crystals with osmolality in 19 samples with pH range 7.0–7.2. Once again only a weak correlation is found. The mean crystal concentrations in urine samples represented on this graph with osmolality higher than 500 ($n = 9$) is nearly twice the corresponding volumes where the urine osmolality was 500 or less ($n = 10$), none the less the difference does not reach statistical significance. Precipitation of calcium phosphate sometimes occurs abruptly in otherwise crystal-free urines. This complicates measurements of the concentrations of crystals of calcium phosphate and may, in part, account for the weak relationship between phosphate crystalluria and overall urine concentration.

pH Effects on Crystalluria

Microscopy Findings

The percentage of urines from stone-formers showing envelope oxalate crystals in various pH bands may be seen in Fig. 9.19a, and the corresponding percentages for calcium phosphate in Fig. 9.19b. Note that there is an apparent decrease in the frequency of observation of envelope oxalate crystals with increasing pH, whilst the reverse appears true for calcium phosphate which is rarely encountered below pH 5.6. At low urinary pH values envelope oxalate crystals are seen in nearly half of the urines (untreated plus post-evaporation) whilst at the highest pH range nearly all samples will show calcium phosphate crystals, whilst envelope oxalate crystals are rarely seen.

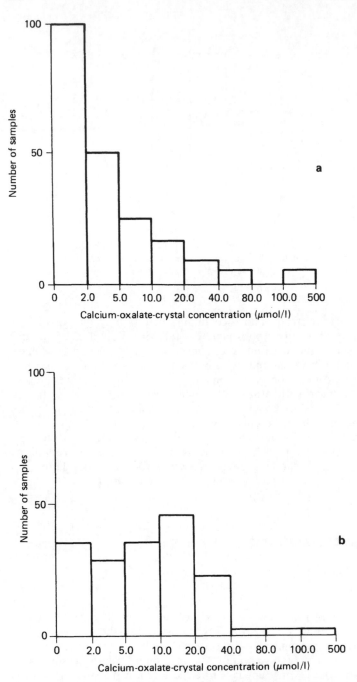

Fig. 9.20. **a** The distribution of concentrations of crystals of calcium oxalate in stone-former urines in pH range 4.8–6.4. **b** The distribution of concentrations of crystals of calcium oxalate in stone-former urines in pH range 6.4–7.6. Here the calcium oxalate is almost always associated with calcium phosphate.

Effect on Crystalluria of Urinary pH

Calcium Oxalate

The concentration of crystals of calcium oxalate in urine may vary by more than 1000:1; for example from 0.2 μmol/l at the lower end to 500 μmol/l at the upper extreme. Figures 9.20a,b show the distribution of concentrations of crystals of calcium oxalate in concentration bands up to 500 μmol/l. It is clear that the distribution is markedly skewed towards the lower end. Hence mean values of oxalate crystal concentrations are also likely to be low but subject to severe distortion by inclusion of even a single sample with high values. This is illustrated in Fig. 9.21 where mean concentration of calcium oxalate is seen to vary with pH. The pH band 5.6–6.0 has been distorted by inclusion of a single sample with an atypical crystal concentration for this pH class. The dotted line indicates the level when this value is removed. Minimal crystalluria occurs at pH 6.0–6.4 in keeping with earlier findings (Robertson 1969).

It is interesting to compare Fig. 9.21 with Fig. 9.19a. While microscopy evidence may suggest that there is a decrease in oxalate crystalluria with rising pH, chemical analysis indicates that there is no such fall. This is evidently a consequence of the presence of an amorphous form of calcium oxalate (or the oxalate anion) in urinary sediments which contain calcium phosphate.

Low concentrations of calcium oxalate crystals are, as already pointed out, a frequent finding (Fig. 9.20a,b). Figures 9.22a,b illustrate how the higher concentrations (ie. above 5 μmol/l) of calcium oxalate crystals are distributed across the range of urine pH values, for untreated and evaporated urines respectively. These figures make clear that higher concentrations of crystals of calcium oxalate are encountered, not at low pH values as microscopy would

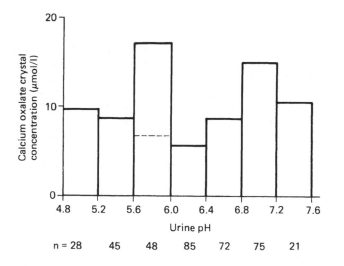

Fig. 9.21. Mean concentration of crystals of calcium oxalate in post-evaporation stone-former urines. The dotted line indicates the result after the exclusion of one atypical value (see text).

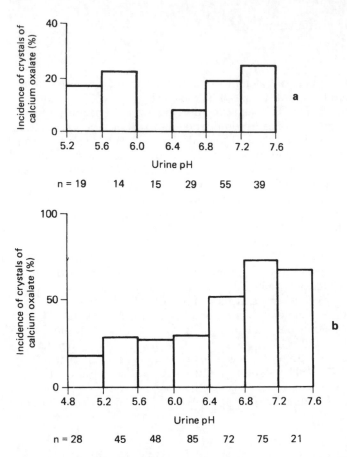

Fig. 9.22. **a** Percentage incidence of concentrations over 5 μmol/l of crystals of calcium oxalate in untreated urines of stone-formers plotted against pH of urine. **b** Percentage incidence of concentrations over 5 μmol/l of crystals of calcium oxalate in evaporated urines of stone-formers plotted against pH of urine.

suggest, but at higher urine pH in association with amorphous calcium phosphate. Note also that the concentration of crystals of calcium oxalate rises between pH 6.4 and 7.6 and that this rise is closely paralleled by the increase in the concentration of crystals of calcium phosphate with increasing pH (Figs. 9.23a,b). The similarity is apparent in both untreated and evaporated urine samples. The distribution of concentration of crystals of calcium oxalate is different therefore from that of calcium phosphate at the lower pH ranges

Fig. 9.23. **a** Mean concentrations of crystals of calcium phosphate in untreated urines ▶ from stone-formers plotted against pH of urine. **b** Mean concentrations of crystals of calcium phosphate in evaporated urines from stone-formers plotted against pH of urine.

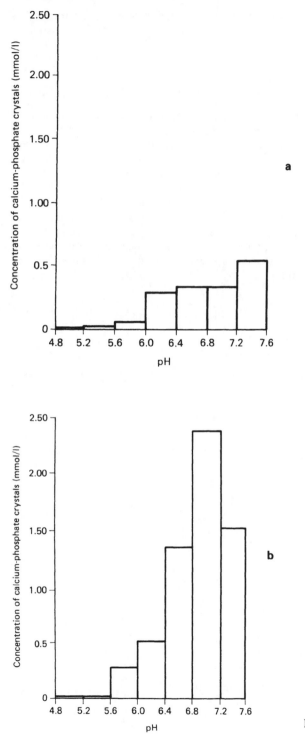

Fig. 9.23

(below pH 6.2) where calcium oxalate manifests in pure form, mainly as envelope-type crystals, and not in conjunction with amorphous calcium phosphate.

Amorphous Particles at low Urinary pH

In a number of instances trace amounts of amorphous particles are seen by light microscopy at the acidic end of the urine pH range (below pH 6.5). These particles usually appear coarser and less clustered than amorphous calcium phosphate. Analysis indicates that these deposits, when not pure calcium oxalate, are relatively very rich in this compound whilst the content of calcium phosphate is close to the lower limit of detection. Figure 9.24 shows a distribution of the concentration of crystals of calcium oxalate of such amorphous deposits. This is seen to match that shown in Fig. 9.20a much more closely than that shown in Fig. 9.20b, and adds weight to the evidence indicating that the deposits are, in the main, an unrecognised amorphous form of calcium oxalate. They are more frequently encountered in evaporated urines but are also seen in untreated specimens.

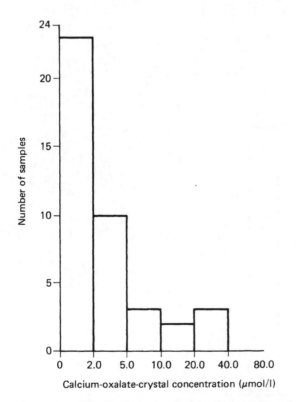

Fig. 9.24. Distribution of concentrations of deposits of amorphous calcium oxalate found in stone-forming and normal subjects, in urine below pH 6.5.

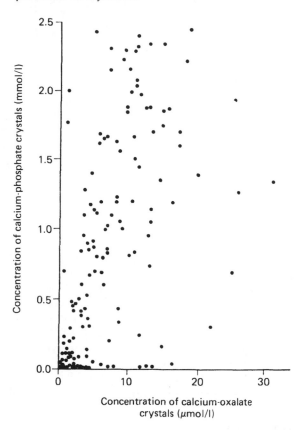

Fig. 9.25. Concentration of crystals of calcium phosphate plotted against corresponding concentrations of crystals of calcium oxalate in urines from stone-formers.

Oxalate Crystals in Association with Amorphous Calcium Phosphate

Comparison of Fig. 9.22a,b with Fig. 9.23a,b suggests a proportionality between the concentrations of calcium phosphate and oxalate crystals. Figure 9.25 shows individual values of the concentration of calcium phosphate crystals plotted against the corresponding concentration of calcium oxalate crystals in each sample. Some scatter of points is present but none the less a discernible correlation between the concentrations of the two crystal types is evident. The distribution of ratios of calcium phosphate to calcium oxalate found in predominately amorphous calcium phosphate precipitates is illustrated in Fig. 9.26. The mean ratio is 185 parts calcium phosphate to 1 of calcium oxalate. The great excess of calcium phosphate does not rule out the possible existence of a calcium oxalate–phosphate complex salt in small amounts. Other possible explanations are the presence of pure amorphous calcium oxalate (or a form where the familiar envelope structure is too small to be resolved by light microscopy). Alternatively, oxalic acid may itself be strongly bound to known adsorbents such as calcium hydroxyapatite present in the phosphate precipitate.

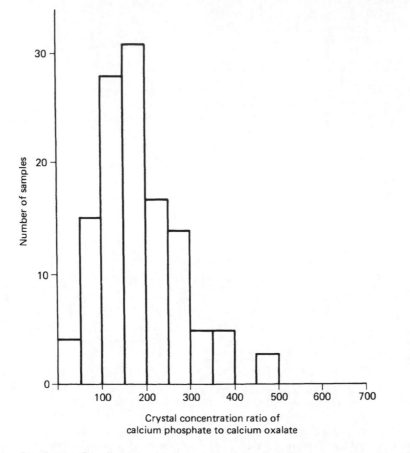

Fig. 9.26. Distribution of ratios of concentrations of crystals of calcium phosphate to calcium oxalate in precipitates predominantly of calcium phosphate from urines of stone-formers. Mean ratio, 185:1; SD, 92.

These findings raise a number of questions in particular the extent to which calcium oxalate coprecipitated with calcium phosphate can contribute to growth of stones. Another query is the degree to which such relatively large precipitates of calcium oxalate might reduce calcium oxalate saturation in urine. Further, what relevance have these findings to phosphate therapy for calcium oxalate stones?

Inhibitors of Calcium Oxalate Crystallisation

Great efforts have been devoted over the years to identifying and evaluating the effects of compounds, both natural and artificial, which may diminish the rate of formation of crystals (Fleisch 1978). Numerous and varied methods for evaluating potential inhibitors have been tried. Results have often been equivocal and sometimes contradictory (Welshman and McGeown 1972).

Hammarsten (1929) was one of the first to recognise the possible value of magnesium in the treatment of calcium oxalate stones. Magnesium forms a soluble complex with oxalate and its ability to lessen calcium oxalate crystalluria has rarely been challenged. Aside from experimental studies (Lyons et al. 1966; Hallson et al. 1982) there is clinical evidence to suggest that magnesium therapy prevents recurrence of calcium oxalate urolithiasis (Dent and Stamp 1970; Johansson et al. 1980).

A first glance at Fig. 9.27 may give the impression that magnesium has no such inhibitory properties. However, once again, such a representation neglects the likely increases in urinary oxalate which may coincide with increases in concentrations of urinary magnesium. An inhibitory effect due to magnesium may then be offset by a correspondingly raised oxalate level. Thus it should be possible to find a magnesium effect when urinary oxalate is fairly constant and where changes in its concentration produce minimal changes in oxalate crystalluria. When urine samples below 0.10 mmol/l are divided into two groups, one containing magnesium above the mean concentration for all the

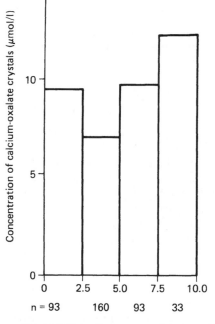

Fig. 9.27. Concentrations of crystals of calcium oxalate plotted against concentrations of urinary magnesium in urines from stone-formers.

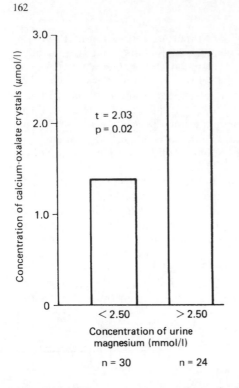

Fig. 9.28. The mean concentrations of calcium oxalate crystals in urine from 54 stone-formers with magnesium concentrations above and below 2.5 mmol/l where the urine contained no more than 0.1 mmol/l oxalate.

samples, and the second group below this level, the corresponding concentrations of crystals of calcium oxalate differ as shown in Fig. 9.28. Nevertheless, as in the preceding figure, increases in urinary magnesium give the impression that crystal formation is promoted.

Citric acid is also widely believed to keep sparingly-soluble calcium salts in solution in urine (Elliot and Eusebio 1956; Sutor et al. 1979; Hallson et al. 1983). Other investigations have disputed this (Welshman and McGeown 1972; Pyrah 1979; Hodgkinson 1980). As with magnesium, the crystal-forming effect of increasing urinary oxalate may obscure any opposing effect of citrate (Fig. 9.29). The raw data in this study are derived from a large number of urine samples in which there is little knowledge or control over the numerous ionic, molecular, macromolecular and solid phase constituents present. It is possible that the cumulative effect of these components, some inhibitory, others promoting crystal growth may be sufficient to conceal any lessening of calcium crystalluria by citrate concentrations within the physiological range. In contrast to this, only in experiments in which urine composition is held constant are the effects of changes in magnesium or citratre concentration likely to be apparent (Hallson et al. 1983). The failure to find an inhibitory effect of citrate in his study involving numerous random urines may serve to show that the

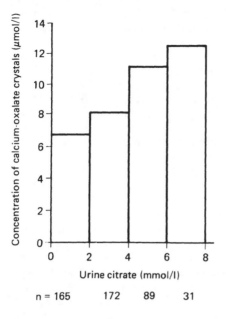

n = 165   172   89   31

Fig. 9.29. Concentrations of crystals of calcium oxalate plotted against concentrations of urinary citrate in urines of stone-formers.

interactions in urine which promote and inhibit crystalluria may be far more complex than envisaged.

Summary

Combining microscopy, urine evaporation and a sensitive assay for oxalate crystals has enabled research into oxalate crystalluria which has not been possible previously.

That the level of urinary oxalate plays a decisive role in promoting calcium oxalate crystalluria is very apparent (Fig. 9.4). Furthermore, when the former exceeds the normal range the rate of calcium oxalate crystallisation increases manyfold. Microscopy shows a similar abrupt increase in the frequency with which envelope oxalate crystals are seen above the top of the normal oxalate range (Fig. 9.6) and these findings must have implications for treatment.

By contrast, urine calcium near the top of normal levels (6 mmol/l) yields concentrations of crystals of oxalate similar to those well within the normal urine oxalate band. Higher urine calcium values, corresponding to hypercalciuria, give concentrations of crystals of calcium oxalate close to those generated by urinary oxalate slightly over the normal levels. Hence, whilst calcium is clearly the lesser contributor to crystalluria, its effects cannot be neglected (Fig. 9.13).

As the concentration of oxalate in urine rises, the frequency with which aggregates are seen under the microscope increases sharply from about 1%

incidence at low concentrations of oxalate to one-third in all urines examined in the hyperoxaluric range (Fig. 9.9). The increasing incidence of aggregates at higher levels of urinary oxalate is accompanied by increasingly frequent sightings of the dumb-bell form of calcium oxalate crystal. The latter are themselves polycrystalline aggregates and this leads one to suspect, therefore, that there is a close connection between the level of urinary oxalate and the capacity of crystals of calcium oxalate to agglomerate. By contrast, it is interesting that the size of the largest aggregates is unaffected by the level of urine oxalate.

Previously, when envelope oxalate crystals had not been observed microscopically in amorphous calcium phosphate, it was (evidently) assumed that such precipitates were oxalate-free acidic or basic salts of calcium phosphate. This does not seem to be true; in fact more oxalate appears coprecipitated with calcium phosphate than occurs in pure form at lower pH values. Whether coprecipitated oxalate is present as a calcium oxalate–phosphate salt, as adsorbed oxalate, or as pure calcium oxalate is not clear, except that it is a fairly constant fraction of the total mass of calcium phosphate. Increasing pH of urine yields larger precipitates of calcium phosphate (Fig. 9.23a, b) and this is accompanied by increasing oxalate crystalluria. There is also some evidence for the occurrence of an amorphous form of calcium oxalate (in the absence of calcium phosphate) at urine pH below 6.5. The implication of these findings in relation to stone-growth justifies further inquiry.

In contrast with the random urines examined in the present study, in vitro tests of potential crystal inhibitors usually employ a crystallising medium of constant composition, where it is possible to increase the concentration of the alleged inhibitor to above physiological levels. No such control over the numerous urinary solutes, colloids and particles is possible in random urines. This, perhaps, serves to underline the potency of unknown constituents and unknown interactions in urine, and to put into perspective the capacity of magnesium and citrate to diminish oxalate crystalluria in widely differing urines.

References

Azoury R, Garside J, Robertson WG (1986) Calcium oxalate precipitation in a flow system: an attempt to simulate the early stages of stone-formation in the renal tubules. J Urol 136: 150–153

Ball WG, Evans G (1932) Diseases of the kidney. Churchill, London

Berg W, Schnapp J-D, Schneider H-J, Hesse A, Hienzsch E (1976) Crystaloptical and spectroscopical findings with calcium oxalate crystals in the urine sediment. Eur Urol 2: 92–97

Catalina F, Cifuentes L (1970) Calcium oxalate: crystallographic analysis in solid aggregates in urinary sediments. Science 169: 183–184

Dent CE, Stamp TCB (1970) Treatment of primary hyperoxaluria. Arch Dis Child 45: 735–745
Dyer R, Nordin BEC (1967) Urine crystals and their relation to stone formation. Nature 215: 751–752
Elliot JS, Eusebio E (1956) Calcium oxalate solubility; the effect of inorganic salts, urea, creatinine and organic salts. Invest Urol 3: 72–76
Finlayson B (1978) Physico-chemical aspects of urolithiasis. Kid Int 13: 344–360
Fleisch H (1978) Inhibitors and promotors of stone formation. Kid Int 13: 361–371
Hallson PC (1977) Studies of urine composition in stone-formers and normal subjects. PhD Thesis. London University
Hallson PC, Rose GA (1976) Crystalluria in normal subjects and in stone formers with and without thiazide and cellulose phosphate treatment. Br J Urol 48: 515–524
Hallson PC, Rose GA (1977) Seasonal variations in urinary crystals. Br J Urol 49: 277–284
Hallson PC, Rose GA (1978) A new urinary test for stone 'activity'. Br J Urol 50: 442–448
Hallson PC, Rose GA (1979) Uromucoids and urinary stone formation. Lancet I: 1000–1002
Hallson PC, Rose GA (1988) Procedure for the measurement of calcium oxalate and phosphate crystals in urine. Br J Urol (In press)
Hallson PC, Rose GA, Sulaiman S (1982) Magnesium reduces calcium oxalate crystal formation in human whole urine. Clin Sci 62: 17–19
Hallson PC, Rose, GA, Sulaiman S (1983) Raising urinary citrate lowers calcium oxalate and calcium phosphate crystal formation in whole urine. Urol Int 38: 179–181
Hammarsten G (1929) On calcium oxalate and its solubility in the presence of inorganic salts with special reference to the occurrence of oxaluria. CR Lab Carls 17: 1–83
Hodgkinson A (1980) Solubility of calcium oxalate in human urine, simulated urine and water. Invest Urol 18: 123–136
Johansson G, Backman U, Danielson G, Fellström B, Ljunghall S, Wikström B (1980) Biochemical and clinical effects of the prophylatic treatment of renal calcium stones with magnesium hydroxide. J Urol 124: 770–774
Kasidas GP, Rose GA (1985) Continuous flow assay for urinary oxalate using immobilised oxalate oxidase. Ann Clin Biochem 22: 412–419
Lyons ES, Borden TA, Ellis JE, Vermeulen CW (1966) Calcium oxalate lithiasis produced by pyridoxine deficiency and inhibition by high magnesium diets. Invest Urol 4: 133–142
Mullen JW (1972) Crystallisation. Butterworth, London
Pyrah LN (1979) Renal calculus. Springer, Berlin
Robertson WG (1969) A method for measuring calcium crystalluria. Clin Chim Acta 26: 105–110
Robertson WG, Nordin BEC (1969) Activity products in urine. In: Hodgkinson A, Nordin BEC (eds) Proceedings of the renal stone research symposium, Leeds 1968. Churchill, London, pp 221–232
Robertson WG, Nordin BEC (1976) Physico-chemical factors governing stone formation. In: Innes-Williams D, Chisholm GD (eds) Scientific foundations of urology, Volume 1, pp 254–267
Robertson WG, Peacock M (1980) The cause of idiopathic calcium stone disease: hypercalciuria or hyperoxaluria? Nephron 26: 105–110
Robertson WG, Peacock M, Nordin BEC (1968) Activity products in stone forming and non-stone forming urine. Clin Sci 34: 579–594
Robertson WG, Peacock M, Nordin BEC (1969) Calcium crystalluria in recurrent renal stone formers. Lancet II: 21–24
Robertson WG, Peacock M, Heyburn PJ, Marshall RW, Williams RE, Clark PB (1979) The significance of mild hyperoxaluria in calcium stone-formation. In: Rose GA, Robertson WG, Watts RWE (eds) Oxalate in human biochemistry and clinical pathology. Proceedings of an International Meeting in London. Wellcome, London, pp 173–180
Robertson WG, Peacock M, Heyburn PJ, Bambach CP (1981) Risk factors in calcium stone disease. In: Brockis JG, Finlayson B (eds) Urinary calculus. PSG, Littleton, Mass. pp 265–273
Rose GA, Sulaiman S (1984) Tamm-Horsfall mucoprotein promotes calcium phosphate crystal formation in whole urine: quantitative studies. Urol Res 12: 217–221
Rushton HG, Spector M, Rodgers AL, Hughson M, Magura CE (1981) Developmental aspects of calcium oxalate tubular deposits and calculi induced by rat kidneys. Invest Urol 19: 52–57
Scurr DS, Robertson WG (1986) Modifiers of calcium oxalate crystallisation in urine II. Studies on their mode of action in an artificial urine. J Urol 136: 128–131
Sutor DJ, Percival JM, Doonan S (1979) Urinary inhibitors of the formation of calcium oxalate. Br J Urol 51: 253–255
Von Sengbusch R, Timmerman A (1957) Das Kristalline calcium oxalat im menschilichen Harn

und seine Berziehung zur oxalatstein-bildung. Urol Int 4: 76–95

Welshman SG, McGeown MG (1972) A quantitative investigation of the effects on the growth of calcium oxalate crystals on potential inhibitors. Br J Urol 44: 677–680

Werness PG, Brown CB, Smith LH, Finlayson B (1985) Equil 2: A basic computer programme for the calculation of urinary saturation. J Urol 134: 1242–1244

Chapter 10

Renal Failure and Transplantation in Primary Hyperoxaluria

M. A. Mansell and R. W. E. Watts

Introduction

Primary hyperoxaluria is a rare inherited metabolic disease in which striking oxalate overproduction is associated with recurrent calcium oxalate urolithiasis leading to end-stage renal failure before the end of the second or third decade. Because conventional methods of renal substitution such as regular dialysis or renal transplantation do not alter the underlying metabolic defect the longterm results of such treatments are far worse in patients with primary hyperoxaluria than in patients whose renal failure is due to other more common causes such as chronic glomerulonephritis or pyelonephritis. This chapter reviews recent developments in the management of patients with primary hyperoxaluria as they approach terminal renal failure, together with the timing and management of renal transplantation and the option of combined synchronous hepatic and renal transplantation.

Clinical Aspects

Full accounts of primary hyperoxaluria in children (chap. 6) and adults (chap. 5) appear elsewhere in this volume. Briefly, the more common form of the disease (type I) is due to deficiency of hepatic peroxisomal alanine:glyoxylate aminotransferase (AGT, EC 2.6.1.44) which causes increased production and urinary excretion of both oxalate and glycollate (Danpure and Jennings 1986) and the clinical severity reflects the degree of AGT deficiency (Danpure et al. 1987). Urinary oxalate is commonly increased to 2–6 mmol/24 h, compared with normal values up to about 0.4 mmol/24 h and levels in patients with secondary hyperoxaluria (eg. after jejuno-ileal bypass) of up to about 0.8 mmol/24 h. Type II patients, of whom only 4 are published (Williams and Smith 1983) have increased urinary excretion of oxalate and L-glycerate with normal glycollate, and other biochemical variants also exist (Yendt and Cohanim 1985).

The first symptoms due to calculus formation commonly occur in childhood and are followed by a progressive decline in renal function, punctuated by acute episodes of renal damage due to obstruction in the upper or lower urinary tract. The management of the illness involves meticulous out-patient follow-up combined with frequent hospitalisation and it is important that these patients can develop a good relationship with the nephrologist and urologist who are, in combination, responsible for their on-going management.

Medical Care (see Table 10.1)

About 30% of all patients with type I primary hyperoxaluria will respond to high doses of pyridoxine (800 mg/day) with a sustained fall in urinary output of oxalate (Gibbs and Watts 1970; Watts et al. 1985a) and the effects of a therapeutic trial, for at least six months, should be evaluated in all patients. Additional aspects of medical management include maintenance of a high intake of fluid and non-specific inhibitors of crystallisation such as magnesium or phosphate, and the control of hypertension and urinary infection. Dehydration should be anticipated where possible and avoided at all costs, with an intravenous infusion being established before even the simplest of surgical procedures. In patients who have passed ureteric calculi on previous occasions the development of ureteric obstruction may be completely silent and a high index of clinical suspicion must always be maintained. Renal ultrasound and diethylenetriaminepentaacetic acid (DTPA) scanning with frusemide are the diagnostic methods of choice to exclude obstruction and repeated intravenous urography should be avoided whenever possible because of the large cumulative radiation dose that these patients may otherwise receive.

Surgical Care (see Table 10.2)

Hitherto these patients have presented formidable problems for the urologist responsible for dealing with their recurrent calculous obstruction. Repeated open surgical removal of stones from the renal pelvis or ureter hastens the progression of renal impairment and the catastrophe of a nephrectomy being required because of a pyonephrosis is by no means unknown. Modern techniques such as percutaneous nephrolithotomy and extracorporeal shock-wave lithotripsy allow many stones to be dealt with noninvasively that would previously have required open operation. The development of nephroscopes and ureteroscopes with laser or electrohydraulic means of stone disintegration also offers the prospect of better preservation of renal function for these

Table 10.1.   Clinical management of primary hyperoxaluria

A   *General measures*
   High fluid intake
   Crystallisation inhibitors
      Magnesium/phosphate
   Pyridoxine
      Effective in 30% (200 mg qds > 6 months)
   Check patient compliance
   Anticipate and avoid dehydration

B   *Are they obstructed?*
   Often clinically silent
   Ultrasound/DTPA diagnosis
   Minimise radiation dose

C   *Other problems*
   Hypertension/urinary infection

patients in the future. Problems still remain, however, with the very large and hard masses of stone presented by calcium oxalate staghorn calculi. These stones will need initial debulking by percutaneous nephrolithotomy before lithotripsy can be undertaken and the resulting debris may well require external drainage via a nephrostomy or internal drainage via a ureteric stent. Such drainage tubes may become rapidly encrusted with oxalate deposits so that they become blocked and difficult to remove without ureteric damage which may lead to the formation of strictures. Although lithotripsy in normal kidneys has little effect on renal function unless obstruction occurs, there is an impression that kidneys with interstitial oxalate deposition may be particularly susceptible to lithotripsy-induced renal damage. It is too soon to say if modern, noninvasive techniques of stone removal will indeed have a significant beneficial effect on the natural history of the disease.

Table 10.2.   Surgical management of primary hyperoxaluria

A   *Principles*
   Conservative and noninvasive
   Preserve renal tissue
   Nephrologist/urologist cooperation

B   *Techniques*
   Percutaneous nephrolithotomy
   Ureteroscopy/nephroscopy
   Laser/electrohydraulic disintegration
   ESWL

C   *Problems*
   Very large stone masses
      May need debulking
      Prophylactic stents
   Ureteric strictures
   ?Lithotripsy-induced renal damage
   Stent encrustation with calcium oxalate

Systemic Oxalosis

Urinary oxalate excretion is by a combination of glomerular filtration and tubular secretion (Cattell et al. 1962; Constable et al. 1979) so that progressive renal impairment leads to a fall in urinary output of oxalate, which then accumulates in the body (Stansfeld et al. 1959). Oxalate deposits in peripheral vessels may cause gangrene, cardiac deposits, cause heart block and sudden death while deposition in bone may cause crippling pain (Gherardi et al. 1980), especially in children on dialysis (Breed et al. 1981; Adams et al. 1982). The deposition of oxalate in the kidney further impairs renal function, so accelerating the process and renal function may decline rapidly from an apparently stable glomerular filtration rate (GFR) of about 30 ml/min to virtually zero within a period of weeks or months. Measurement of blood urea and creatinine gives a relatively insensitive guide to the progression of renal failure and sequential measurement of GFR, by creatinine clearance or, ideally, ^{51}Cr-EDTA, is necessary. Watts et al. (1983) have described an isotopic technique which allows measurement of plasma oxalate, the metabolic pool size of oxalate and the accretion rate of tissue oxalate and thus gives a more complete picture of the dynamic changes which occur in the handling of oxalate by the body as renal failure develops. The problems of measuring plasma oxalate are discussed elsewhere in this volume (chapter 4), although an accurate biochemical method now exists which will undoubtedly be widely used in the future for sequential studies (Kasidas and Rose 1986).

The development of systemic oxalate deposits is associated with the occurrence of serious medical complications and indicates that the ideal time has already passed for discussion and planning of renal substitution therapy. Recent studies by Morgan et al. (1987) in patients with primary hyperoxaluria have shown that values for plasma oxalate, metabolic pool size and tissue accretion rate start to increase as the GFR falls below about 50–60 ml/min/1.73 m^2. This is in contrast to patients with chronic renal failure due to other causes in whom the changes in oxalate parameters only become detectable with a GFR of less than 20–25 ml/min/1.73 m^2 and are of far lesser degree than in patients with primary hyperoxaluria. Clearly, plans for dialysis or transplantation need to be made when the GFR is only slightly reduced, with a view to initiation when the GFR is no less than 25 ml/min/1.73 m^2 so that oxalate accumulation can be minimised. This level of renal function is, of course, far greater than the value at which renal replacement therapy for standard biochemical criteria is necessary.

Peritoneal Dialysis and Haemodialysis

It is well-established that attempts to treat end-stage renal failure in patients with primary hyperoxaluria by conventional peritoneal dialysis or haemodialysis schedules are unsuccessful because there is poor oxalate removal relative to its production (Zarembski et al. 1969; Arbus and Sniderman 1974; Fayemi et al. 1979; Williams and Smith 1983). Oxalosis causes problems from heart block,

Table 10.3. Oxalate removal by dialysis procedures

A   *Properties of oxalate*
   Molecular weight 88 daltons
   Double negative charge
   86% ultrafilterable in plasma
   50% reduction in plasma oxalate with 6-h haemodialysis
   ? poor removal across PAN membrane

B   *Results of a 6-h haemodialysis (1 sq M cuprophane dialyser)*
   Plasma oxalate predialysis 111 μmol/l
   Oxalate dialysance = 52 ml/min
   Oxalate removal = 2.05 mmol/6 h
   Oxalate production = 3.87 mmol/24 h
   Tissue oxalate accretion:-
      Daily dialysis = 1.82 mmol/24 h
      3 dialyses/week = 3 mmol/24 h

C   CAPD
   Peritoneal clearance = 5 ml/min
   Peritoneal removal = 0.77 mmol/24 h
   Tissue oxalate accretion = 3.1 mmol/24 h

Data from Watts et al. (1984)

bone pain and peripheral gangrene so that the quality of life for these patients on dialysis is generally miserable. Oxalate is a small molecule (MW 88), intermediate in weight between urea and creatinine, so that the dialysance across a standard cuprophan membrane is good; the problem arises because of the striking degree of overproduction in primary hyperoxaluria (see Table 10.3). A 6-h dialysis will produce a fall in plasma oxalate of about 50% (Boer et al. 1984), although there is a suggestion that use of the new, highly-permeable polyacrylonitrile or polycarbonate membranes which contain fixed negative charges could be less effective, because of the double negative charge on the oxalate ion (Rose G A, personal communication). Watts et al. (1984) studied a 6-h haemodialysis with a 1 m² cuprophan dialyser and found an oxalate dialysance of 52 ml/min, equivalent to removal of 2.05-mmol oxalate during the dialysis period. In the patient studied the rate of production of oxalate was 3.87 mmol/24 h, corresponding to a tissue-oxalate accretion rate of 1.72 mmol/24 h with daily dialysis or 3 mmol/24 h with a standard thrice-weekly regime. In this study results were even worse with chronic ambulatory peritoneal dialysis; the peritoneal clearance was only 5 ml/min, equivalent to oxalate removal of 0.77 mmol/24 h and a tissue accretion rate of 3.1 mmol/24 h. Although others have reported values for oxalate removal of 3 mmol/4-h dialysis (Ramsay and Reed 1984) and 5 mmol/4.5 h haemodialysis/haemoper-fusion (Ahmed and Hatch 1985) there are major uncertainties about the accuracy of the methods used for oxalate measurement in these studies. Our unpublished exprience with haemodiafiltration suggests that this method has no particular advantages in terms of oxalate removal and, as with other techniques, the problem is of overproduction.

It seems clear that currently-available dialysis methods offer no longterm solution for the patient approaching end-stage renal failure from primary hyperoxaluria. In a typical case high-efficiency haemodialysis for 6–8 h/day would be required just to keep pace with oxalate production, let alone remove systemic oxalate deposits, and this would be clinically intolerable for more than a short period. Nevertheless, dialytic removal of oxalate does have an important part to play in the management of the renal transplant recipient and this is considered in the next section. First, however, some general therapeutic principles are outlined in Table 10.4.

Table 10.4. Treatments available for primary hyperoxaluria

A   *Conservative management*
    Fluids/pyridoxine/magnesium/phosphate
    Relief of obstruction/infection/hypertension
    Sequential studies of renal function and oxalate dynamics
    Make plans for the future

B   *Renal transplantation alone*
    May give longterm graft survival
    Metabolic defect unaltered
    Lesser surgical procedure

C · *Hepatic/renal transplantation*
    Major surgical procedure
    Complete correction of metabolic defect
    No longterm results as yet

D   *Hepatic transplantation alone*
    Must be before renal failure
    Is it justified?

Renal Transplantation

The consensus view of the American College of Surgeons and National Institutes of Health in 1975 was that "oxalosis is a form of chronic renal failure that is unsuitable for treatment by renal transplantation" (Wilson 1975) and most of the evidence supports that conclusion (Klauwers et al. 1969; Deodhar et al. 1969; Williams and Smith 1983; Chesney et al. 1984; Vanrenterghem et al. 1984), although recently a number of exceptions have appeared (David et al. 1983; Whelchel et al. 1983; Scheinman et al. 1984; Watts et al. 1987a).

The main problem following transplantation in patients with primary hyperoxaluria is of oxalate deposition within the parenchyma of the graft with tubular obstruction and irreversible damage. Plasma oxalate levels in these patients may be greater than 150 μmol/l, yielding a [calcium] [oxalate] concentration product many times the solubility product. Moreover, there is also a suggestion that renal tissue which is the site of ischaemic or immunological injury may be particularly susceptible to the deposition of oxalate. If good renal excretory function is obtained then extreme hyperoxaluria results from the greatly raised level of plasma oxalate and exchangeable metabolic pool, and calculus formation with the risk that obstruction to drainage of the transplant may develop very rapidly indeed, within a period of weeks or months. If the timing of transplantation has been unduly delayed the complications of systemic oxalosis add to the risks of the procedure. Intracardiac deposits may cause heart block and other serious dysrhythmias while arterial deposits lead to peripheral vascular insufficiency and particular problems with vascular access. We have also seen a ureteric stent become so heavily encrusted with oxalate within six weeks of transplantation that it became blocked and impossible to remove via the cystoscopic route.

Based on our experience and that described by Scheinman et al. (1984) we would suggest that the following points are important in the management of patients with primary hyperoxaluria receiving renal transplants.

1. Sequential studies of renal function and oxalate dynamics are needed so that the development of the terminal phase can be anticipated and plans for transplantation made well in advance

2. Intensive haemodialysis will be required in the pretransplant period to reduce the degree of oxalate accumulation, and adequate vascular access capable of delivering high blood-flows must be established in good time. A typical schedule would consist of daily haemodialysis for 6–8 h with a 1.5 m^2 dialyser and the problems of patient compliance with this harsh regime must be anticipated

3. If a suitable live-related donor exists then transplantation can be undertaken while a useful degree of residual renal function is still present and immediate function can be anticipated posttransplant. A good HLA match will reduce the likelihood of rejection episodes accelerating graft oxalosis because of impaired function although, clearly, the presence of a forme fruste of the disease in a sibling donor must be rigorously excluded

4. Immunosuppression should be with cyclosporin A and low-dose prednisolone and the recipient should have been previously transfused to optimise graft survival. The use of azathioprine carries a greater risk of ureteric leaks and wound infections and should be avoided when possible. A ureteric stent will protect the vesicoureteric anastamosis and, perhaps, reduce the risk of ureteric calculous obstruction although it should be removed as quickly as possible because of the risk of encrustation with oxalate

5. Postoperative diuresis should be encouraged by adequate hydration and the administration of agents such as frusemide, dopamine or prostacyclin

6. Intravenous neutral phosphate (2 g/day of phosphorus) and magnesium (10 mmol/day) may reduce oxalate crystallisation and should be continued until oral supplements can be taken satisfactorily. If the patient has been shown to be responsive to pyridoxine full doses (800 mg/day) should be continued longterm.

7. Even if good immediate function is established with a diuresis of 5–10 l/day, high-efficiency haemodialysis should continue for at least 7–10 days postoperatively. Even a modest decline in renal function, for example due to rejection or cyclosporin toxicity, should be covered by intensive haemodialysis. The distinction between graft oxalosis, rejection and cyclosporin toxicity may be impossible on clinical grounds and there should be a low threshold for performing a transplant biopsy

Scheinman et al. (1984) reported success, with follow-up for 1–8 yr in 7 of 11 transplant recipients, 10 of which were live-related, with an average oxalate output of around 2 mmol/24 h. Watts et al. (1987a) reported 6 patients, with success in 3 (1 cadaver, 2 live-related) whose disease would seem to have been of greater severity, as judged by an average production rate of oxalate of about 5 mmol/day. Both groups used the approach to transplantation outlined above and agreed that the failures occurred mainly in patients who had been in established end-stage renal failure for a long time and/or had not been treated with an aggressive dialysis regime. A live-related donor is not a prerequisite for success.

Combined Hepatic and Renal Transplantation

The combined procedure allows correction of the underlying metabolic defect and the resulting end-organ damage, providing the native liver is removed. The first recorded attempt was in a patient who was developing systemic oxalosis in whom a cadaveric renal transplant had failed due to a massive deposition of oxalate in the graft (Watts et al. 1985b). The patient succumbed to infection from cytomegalovirus (CMV) eight weeks postoperatively although oxalate overproduction had been corrected by the transplanted liver and there was some evidence that systemic oxalate deposits were being mobilised.

The first success was recently reported (Watts et al. 1987b) in a patient who had previously undergone two renal transplants; the first, from his father, had been lost from a combination of cyclosporin toxicity and calculous obstruction and the second from rejection due to discontinuation of immunosuppression following CMV infection. Ten months after the combined transplant he has a plasma creatinine of 140 μmol/l with a creatinine clearance of 74 ml/min. The plasma oxalate is 5.2 μmol/l and the urinary oxalate is falling, with a most recent value of 0.83 mmol/24 h. He is in full-time employment and recently attended the transplant Olympics.

Conclusion

There have been major advances in the investigation and management of patients with primary hyperoxaluria in recent years. Improved results are now being obtained following renal transplantation so that the procedure is justified in selected patients. The hepatic and renal lesions have been successfully corrected by combined synchronous transplantation, although the follow-up period is still limited. Although there may be immunological advantages to using the same donor for the combined transplant it represents a major procedure; the possibility of correction of the metabolic defect by liver transplantation with renal transplantation at a later date also deserves exploration.

References

Adams MD, Carrera GF, Johnson RP, Latorraca R, Lemann J (1982) Calcium oxalate crystal induced bone disease. Am J Kid Dis 5: 294–299

Ahmed S, Hatch M (1985) Hyperoxalemia in renal failure and the role of hemoperfusion and hemodialysis in primary oxalosis. Nephron 41: 235–240

Arbus GS, Sniderman S (1974) Oxalosis with peripheral gangrene. Arch Pathol Lab Med 97: 107–110

Boer P, van Leersam L, Hene RJ, Mees EJD (1984) Plasma oxalate concentration in chronic renal disease. Am J Kid Dis 4: 118–122

Breed A, Chesney R, Friedman A et al. (1981) Oxalosis-induced bone disease: A complication of prolonged hemodialysis and transplantation in primary oxalosis. J Bone Joint Surg [Am] 63: 310–316

Cattell WR, Spencer AG, Taylor GW, Watts RWE (1962) The mechanism for renal excretion of oxalate in the dog. Clin Sci 22: 43–51

Chesney RW, Friedman AL, Breed AL, Adams ND, Lemann J (1984) Renal transplantation in primary oxaluria. J Pediatr 104: 322–323

Constable AR, Joekes AM, Kasidas GP, O'Regan P, Rose GA (1979) Plasma level and renal clearance of oxalate in normal subjects and in patients with primary hyperoxaluria or chronic renal failure or both. Clin Sci 56: 299–304

Danpure CJ, Jennings PR (1986) Peroxisomal alanine:glyoxylate aminotransferase deficiency in primary hyperoxaluria type I. FEBS Lett 201: 20–24

Danpure CJ, Jennings PR, Watts RWE (1987) Enzymological diagnosis of primary hyperoxaluria type I by measurement of hepatic alanine-glyoxylate aminotransferase activity. Lancet I: 289–291

David DS, Cheigh JS, Stenbzel KH, Rubin AC (1983) Successful renal transplantation in a patient with primary hyperoxaluria. Transplant Proc 15: 2168–2171

Deodhar SD, Tung KSK, Zulke V, Nakamoto S (1969) Renal homotransplantation in a patient with primary familial oxalosis. Arch Pathol Lab Med 87: 118–124

Fayemi A, Ali M, Bruan E (1979) Oxalosis in hemodialysis patients. Arch Pathol Lab Med 103: 58–62

Gherardi G, Poggi A, Sisca S et al. (1980) Bone oxalosis and renal osteodystrophy. Ann Intern Med 90: 777–779

Gibbs DA, Watts RWE (1970) The action of pyridoxine in primary hyperoxaluria. Clin Sci 38: 277–286

Kasidas GP, Rose GA (1986) Measurement of plasma oxalate in healthy subjects and in patients with chronic renal failure using immobilised oxalate oxidase. Clin Chim Acta 154: 49–58

Klauwers J, Wolff PL, Cohn R (1969) Failure of renal transplantation in primary oxalosis. JAMA 209: 551–554

Morgan SH, Purkiss P, Watts RWE, Mansell MA (1987) Oxalate dynamics in chronic renal failure. Comparison with normal subjects and patients with primary hyperoxaluria. Nephron 46: 253–257

Ramsay AG, Reed RG (1984) Oxalate removal by hemodialysis in end-stage renal disease. Am J Kid Dis 4: 123–127

Scheinman J, Najarian JS, Mauer SM (1984) Successful strategies for renal transplantation in primary oxalosis. Kidney Int 25: 804–811

Stansfeld A, Scowen EF, Watts RWE (1959) Oxalosis and primary hyperoxaluria. J Pathol Bacteriol 77: 195–205

Vanrentergham Y, Vandamme B, Lernt T, Michielsen P (1984) Severe vascular complications in oxalosis after successful cadaveric kidney transplantation. Transplantation 38: 93–95

Watts RWE, Veall N, Purkiss P (1983) Sequential studies of oxalate dynamics in primary hyperoxaluria. Clin Sci 65: 627–633

Watts RWE, Veall N, Purkiss P (1984) Oxalate dynamics and removal rates during haemodialysis and peritoneal dialysis in patients with primary hyperoxaluria and severe renal failure. Clin Sci 66: 591–597

Watts RWE, Veall N, Purkiss P, Mansell MA, Haywood EF (1985a) THe effect of pyridoxine on oxalate dynamics in three cases of primary hyperoxaluria (with glycollic aciduria). Clin Sci 69: 87–90

Watts RWE, Calne RY, Williams R et al. (1985b) Primary hyperoxaluria (type I): attempted treatment by combined hepatic and renal transplantation. Q J Med 57: 697–703

Watts RWE, Morgan SH, Purkiss P, Mansell MA, Baker LRI, Brown CB (1987a) Timing of
    renal transplantation in the management of pyridoxine resistant type I primary hyperoxaluria.
    Transplantation (in press)
Watts RWE, Calne RY, Rolles K et al. (1987b) Successful treatment of primary hyperoxaluria
    type I by combined hepatic and renal transplantation. Lancet II: 474–475
Whelchel JD, Alison DV, Luke RG, Curtis J, Dietheim AC (1983) Successful renal transplantation
    in hyperoxaluria. A report of two cases. Transplantation 35: 161–164
Williams HE, Smith LH (1983) Primary hyperoxaluria. In: Stanbury JB, Wyngaarden JB,
    Fredrickson DS, Goldstein JL, Brown MS (eds) The metabolic basis of inherited disease. 5th
    ed. McGraw-Hill, New York, pp 204–208
Wilson RE (1975) A report from the ACS/NIH renal transplant registry. Renal transplantation
    in congenital and metabolic disease. JAMA 232: 148–153
Yendt ER, Cohanim M (1985) Response to a physiological dose of pyridoxine in type I primary
    hyperoxaluria. New Engl J Med 312: 953–957
Zarembski PM, Rosen SM, Hodgkinson A (1969) Dialysis in the treatment of primary
    hyperoxaluria. Br J Urol 41: 530–533

Chapter 11

Vitamin B6 Metabolism in Relation to Metabolic Hyperoxaluria

P. K. S. Liu and G. A. Rose

Introduction

Structure and Function of Vitamin B6

Vitamin B6 is the family name for a group of closely related compounds derived from 3-hydroxy-2-methyl pyridine and exhibiting the biological activity of pyridoxine. Shown in Fig. 11.1 are the members of this family which includes pyridoxine (PN), pyridoxal (PL), pyridoxamine (PM), their respective 5'-phosphate esters, and the dead-end metabolite, 4-pyridoxic acid (4-PA).

The most important and biologically active form of vitamin B6 in blood plasma is pyridoxal phosphate. It functions as a co-factor for many enzymes involved in the metabolism of amino acids. The three main types of reaction which require pyridoxal phosphate are transamination, decarboxylation and racemisation of amino acids.

Vitamin B6 and Hyperoxaluria

It has been shown (Gershoff et al. 1959) that when cats were fed on a diet deficient in vitamin B6 nephrocalcinosis and hyperoxaluria resulted. Similarly, when human subjects were made vitamin-B6 deficient by replacing the vitamin with desoxypyridoxine in their diet, hyperoxaluria also resulted (Faber et al. 1963). Since these initial studies other research workers have confirmed the observation that vitamin-B6 deficiency leads to increased urinary oxalate in both man and animals (Gershoff 1964; Ribaya and Gershoff 1979).

VITAMIN B6 COMPOUNDS

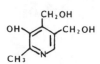

PYRIDOXINE (PN)

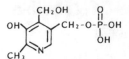

PYRIDOXINE PHOSPHATE (PNP)

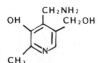

PYRIDOXAL (PL)

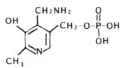

PYRIDOXAL PHOSPHATE (PLP)

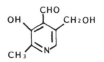

PYRIDOXAMINE (PM)

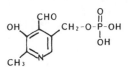

PYRIDOXAMINE PHOSPHATE (PMP)

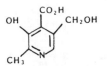

PYRIDOXIC ACID (PA)

Fig. 11.1. Structural formulae of pyridoxine and its known metabolic products.

Type 1 primary hyperoxaluria is an inborn error of metabolism in which there is a deficiency of the peroxisomal alanine: glyoxylate amino transferase leading to excess urinary oxalate and glycollate (see Chap. 5). It was found (Gibbs and Watts 1970) that some cases responded to pyridoxine with falls in the affected metabolites to normal. However, whereas the normal requirement of pyridoxine is thought to be about 5 mg/day, doses of up to 600 mg/day were required to induce these biochemical remissions. Rose (1988) reported that about one-third of cases of type 1 primary hyperoxaluria responded to pyridoxine therapy, another third responded partially or temporarily, while the remaining third showed no response at all. Likewise, in the newly described condition of mild metabolic hyperoxaluria (see Chap. 8), a similar spectrum of response to pyridoxine has been found.

These links between pyridoxine and hyperoxaluria seem to indicate that a study of pyridoxine metabolism in normal subjects and patients with these types of hyperoxaluria might be worthwhile. However, methods of studying pyridoxine metabolism have been poor.

Methods for Studying Pyridoxine Metabolism

Analytical Problems

As shown in Fig. 11.1, pyridoxine is really a family of at least seven structurally-related compounds. The main excretion product is pyridoxic acid which is biologically not a vitamin. The remaining six compounds are all active forms of vitamin B6 and therefore require to be identified in biological samples. However, they vary in nature from acidic to basic and are present in plasma in concentrations measured in ng/ml or less. Therefore the task of measuring them all is quite formidable.

Present Methods Available

A wide range of methods has been developed for measuring one or more forms of the vitamin, including microbiological assays (Polansky 1981; Miller and Edwards 1981), various enzyme assays (Hamfelt 1967; Harskell and Snell 1972; Suelter et al. 1975; Haskell 1981; Lumeng et al. 1981; Yang et al. 1981; Camp et al. 1983; Shin et al. 1983; Lumeng et al. 1984; Singkamani et al. 1986), chemical assays (Takanashi and Tamura 1970; Adams 1979; Dakshinamurti and Chauhan 1981), radioimmunoassays (Thanassi and Cidlowski 1980) and HPLC procedures (Morita and Mizuno 1980; Gregory and Kirk 1981; Gregory et al. 1981; Vanderslice et al. 1981). However, only HPLC procedures have the ability to measure all the substances shown in Fig. 11.1 and we have therefore concentrated on them.

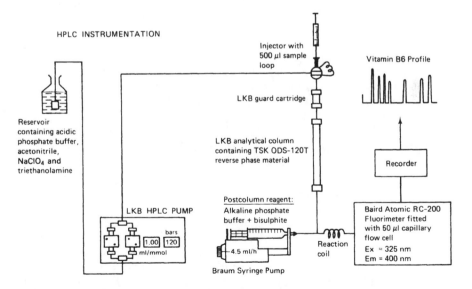

Fig. 11.2. Diagrammatic representation of HPLC system for measuring plasma pyridoxine and its metabolic products.

A Simple HPLC Method for Studying Vitamin B6 Metabolism

The method developed in these laboratories is being published in detail elsewhere (Liu and Rose 1988). A short description only, therefore, is given here, but the method is illustrated in Fig. 11.2. It is a much simplified reverse-phase isocratic procedure with post-column reaction with bisulphite (Coburn and Mahuren 1983). The pump is an LKB model 2150 with titanium parts and the detector is a Baird Atomic spectrofluorimeter (Fluoricord RC200) with a 50 μl capillary flow cell. Excitation and emission wavelengths are 325 and 400 nm respectively. The eluant is NaH_2PO_4 buffer (0.075 M) containing 0.075-M $NaClO_4$, 0.85% acetonitrile and 0.5% triethanolamine and adjusted to pH 3.38 with concentrated $HClO_4$. Immediately after collection, the blood specimens are centrifuged and the plasma separated and either used fresh or quickly frozen at −20°C. Before injection on to the HPLC column, the proteins in the plasma samples are precipitated with 4-M perchloric acid, brought to pH 3.1–3.7 and diluted three- to six-fold and filtered through a millipore Millex-HV filter.

Apotryptophanase Assay

The apotryptophanase assay reported here is a modification of the procedure of Haskell and Snell (1972). In this assay, pyruvate generated by the apotryptophanase reaction is coupled to NADH in the presence of lactate dehydrogenase to form lactate and NAD. The decrease in absorbance at 340 nm due to utilisation of NADH is recorded directly by an LKB reaction rate analyser. Full details of this assay will be published elsewhere.

Urinary Pyridoxic Acid Assay

For urinary 4-pyridoxic acid assays, a disodium hydrogen phosphate buffer (0.05 M, pH 7.00) containing 0.05-M $NaClO_4$, 0.05% triethanolamine and 5.6% acetonitrile is used. Excitation and emission wavelengths are 320 and 420 nm respectively. Sample preparation is very simple and only requires dilution with eluent (from 100–1000 fold depending on concentration).

Evaluation of Plasma Vitamin B6 Methods

Separation of Vitamin B6 Standards

With the HPLC procedure described above, all seven known forms of vitamin B6 were clearly resolved down to baseline as shown in Fig. 11.3. The elution time required for a complete B6 profile in this particular run is about 23 min. If the flow rate is increased to 1.1 ml/min, a complete run can be performed in less than 18 min with very little loss in resolution. Each peak in the chromatogram represents 2.6 ng apart from PM, PMP and PA where the quantity is 1.3 ng.

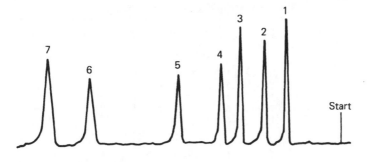

Fig. 11.3. HPLC chromatogram of a mixture of 7 standards. Identifications are as follows: 1, PMP (1.3 ng); 2, PM (1.3 ng); 3, PNP (2.6 ng): 4, PLP (2.6 ng); 5, PL (2.6 ng); 6, PN (2.6 ng); 7, PA (1.3 ng).

Calibration of HPLC System

The system is calibrated by injecting known quantities of B6 and measuring the resulting peak height. A typical calibration curve for pyridoxal phosphate is shown in Fig. 11.4. As illustrated, a linear relationship between peak height and quantity of vitamer injected is clearly seen. Other B6 compounds were calibrated in an identical manner. The sensitivity of the HPLC procedure using the Baird RC200 fluorimeter for the various B6 compounds is about 0.5 ng.

Chemical Recovery of B6 Vitamers from Plasma

To determine the accuracy of the HPLC procedure for measuring plasma-B6 compounds, aliquots of pooled plasma were spiked with a known quantity of B6 (usually 40 µg/l plasma) before perchloric acid precipitation. The recoveries

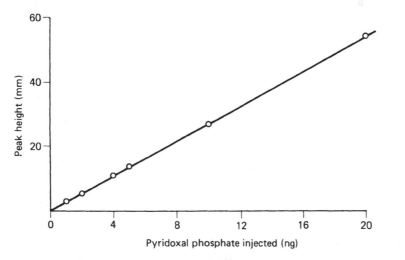

Fig. 11.4. Calibration curve for PLP determined by HPLC.

Table 11.1.   Recoveries of spiked B6 compounds from plasma (HPLC)

Pooled plasma	%PL	%PLP	%PM	%PMP	%PN	%PNP	%4-PA
1	87.0	89.3	97.3	100.0	87.0	97.6	87.1
2	78.3	89.3	94.6	95.4	91.3	90.2	87.1
3	87.0	96.4	97.3	100.0	91.3	97.6	90.3
4	82.6	92.9	97.3	100.0	95.7	95.1	90.3
5	82.6	92.9	94.6	95.4	91.3	92.7	87.1
6	87.0	89.3	94.6	100.0	91.3	92.7	90.3
7	87.0	89.3	94.6	97.7	91.3	92.7	87.1
8	87.0	96.4	97.3	97.7	91.3	95.1	89.9
9	87.0	96.4	94.6	97.7	91.3	97.6	83.9
Mean recoveries	85.1	92.5	95.8	98.2	91.3	94.6	87.5

of the individual B6 compounds are reported in Table 11.1 where it can be clearly seen that they varied from 85 to 98%. Although a satisfactory mean recovery was obtained for plasma PLP in this study, the extraction of the compound by perchloric acid proved more variable than other B6 forms as reported by others (Singkamani et al. 1986) probably due to PLP's protein-binding properties.

Precision of HPLC Assay for Vitamin B6

The within-run precision of the HPLC assay for the various B6 compounds was good (overall CV better than 4%) and is reported in Table 11.2.

Table 11.2.   Precision of nine consecutive B6 assays by HPLC

B6 Vitamer	Coefficient of Variation
Pyridoxal	3.7%
Pyridoxal Phosphate	3.5%
Pyridoxamine	1.5%
Pyridoxamine Phosphate	1.9%
Pyridoxine	2.4%
Pyridoxine Phosphate	2.8%
4-Pyridoxic Acid	2.9%

Correlation between HPLC and Enzymatic Method for Plasma PLP

Fifty plasma PLP levels from pyridoxine-supplemented and un-supplemented individuals were measured by the HPLC and apotryptophanase methods and results compared over a concentration range of 10–200 µg/l. Good agreement was obtained for plasma PLP over this wide concentration range by the two independent methods (R = 0.94; Y = 0.87x − 4.16).

Correlation between Two Methods for Pyridoxic Acid

In a similar correlation study, 36 urinary 4-pyridoxic acid levels up to 800 μmol/l were measured by both the HPLC and a routine laboratory method (Woodring et al. 1964) and the results compared. Again, a good agreement was obtained by the two methods (R = 0.97; Y = 0.93x + 4.89).

Physiology of Pyridoxine

Diurnal Rhythm of Plasma PLP in Normal Subjects

Seven normal individuals who had received no pyridoxine supplement were studied under normal dietary and working conditions. Venous blood samples were withdrawn at 9.30 am, 12.15 pm, 2.30 pm, 4.30 pm and 8.00 pm. Plasma PLP was measured either by apotryptophanase or HPLC methods and the results are shown in Fig. 11.5. The mean values at the various times were 19.3, 18.1, 18.0, 18.1 and 18.4 ng/ml respectively. Hence there appears to be no significant variation in plasma PLP levels during the day when individuals are not taking vitamin-B6 supplements.

A normal subject took 100 mg pyridoxine/day in four divided doses for one week. On the seventh day, 25 mg doses were taken at 8.40 am and 12.45 pm and on this day plasma levels of PLP at 10.05 am, 11.05 am, 12.10 pm and

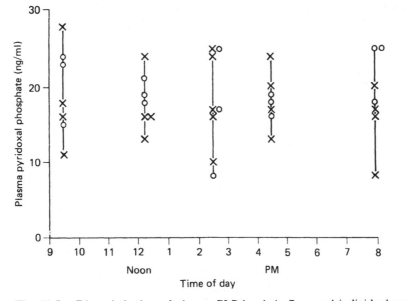

Fig. 11.5. Diurnal rhythm of plasma PLP levels in 7 normal individuals receiving no vitamin supplements. Crosses, assays performed by HPLC; open circles, assays performed by apotryptophanase method.

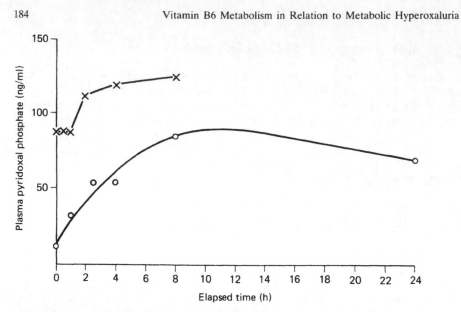

Fig. 11.6. Plasma PLP levels in 2 individuals each of whom received a 200-mg dose of pyridoxine by mouth at time zero. The lower curve (open circles) is from a normal subject who had received no other vitamin supplements. The upper curve (crosses) is from a patient with mild metabolic hyperoxaluria in remission due to longterm doses of pyridoxine (20 mg/day).

1.10 pm were 124, 116, 130 and 125 ng/ml respectively. These values are approximately 6.8 times greater than the basal levels without pyridoxine administration, but there was no significant immediate effect of the dose.

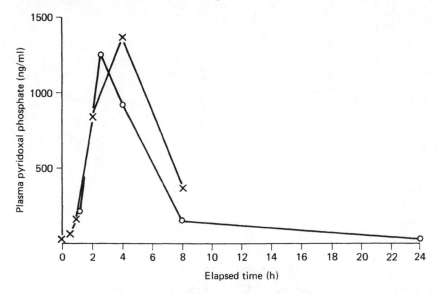

Fig. 11.7. Plasma PL levels following single 200-mg doses of pyridoxine (from the same blood samples used for Fig. 11.6).

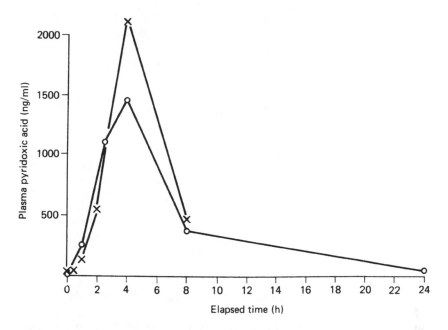

Fig. 11.8. Plasma 4-PA levels following single 200-mg doses of pyridoxine (from the same blood samples used for Fig. 11.6).

Effect of a Single Dose of 200 mg Pyridoxine

One normal individual and one patient with mild metabolic hyperoxaluria who had responded completely to 20 mg/day of pyridoxine were each given single doses of 200 mg pyridoxine while pursuing normal activities. The effects upon plasma PLP levels are shown in Fig. 11.6. In both cases there were considerable rises in plasma PLP levels which appeared to peak at about 6–10 h after the dose.

Figures 11.7 and 11.8 show the plasma PL and 4-PA levels arising from the same blood samples. It is seen that the peak levels for PL are about tenfold higher than those for PLP, but the peaks are reached after only 2–4 h. In the case of 4-PA (Fig. 11.8) peak levels are even higher than those for PL and are attained at about 4 h after the dose of pyridoxine.

In the case of plasma pyridoxine levels (not shown) peak values were reached by about 1 h after dose of pyridoxine and levels were undetectable by 8 h.

Reference Plasma PLP Values

Twenty plasma PLP values from a group of 10 apparently-healthy individuals were determind by the HPLC assay. The reference range was found to be 9–28 µg/l plasma, with a mean of 14.9 µg/l and SD 4.48. These values are in good agreement with both the apotryptophanase assay and those already reported in the literature (Vanderslice et al. 1981; Camp et al. 1983; Coburn and Mahuren, 1983; Singkamani et al. 1986).

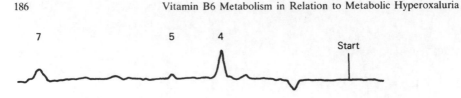

Fig. 11.9. HPLC chromatogram of pyridoxine metabolites in a normal subject taking no vitamin supplements. The main peaks (numbered as in Fig. 11.3) are: 4, PLP; 5, PL; 7, PA.

Plasma B6 Profile from a Normal Subject

A typical B6 chromatogram obtained from the blood plasma of an apparently healthy volunteer taking no vitamin supplements is reported in Fig. 11.9. Using the Baird Fluorimeter at almost maximum sensitivity, the major peaks observed in normal plasma were PLP and PA with PL just detectable. No other B6 forms were detected. The outstanding feature of this HPLC assay is that very little or no interference is seen in plasma-treated samples.

Plasma B6 Profiles from Individuals on Pyridoxine Supplements

Using our newly developed HPLC method for vitamin-B6 analysis, we report here the metabolic profiles obtained from individuals on pyridoxine supplementation. Reported in Fig. 11.10 is the plasma-B6 profile from a patient with primary hyperoxaluria on 800 mg PN/day. In addition to the major

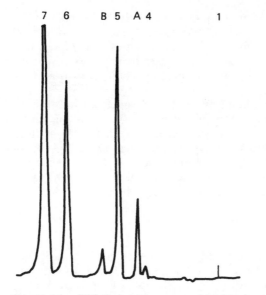

Fig. 11.10. HPLC chromatogram of pyridoxine metabolites in a normal subject taking pyridoxine (800 mg/day), 1-h after administration of pyridoxine. Peak identifications are as for Fig. 11.3: 1, PMP; 4, PLP; 5, PL; 6, PN; 7, PA. Note 2 additional peaks identified as A and B.

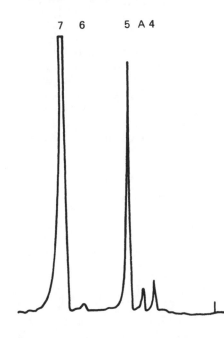

Fig. 11.11. HPLC chromatogram of pyridoxine metabolites in a normal subject taking pyridoxine (800 mg/day) 4 h after the last dose. Identifications as in Fig. 11.3: 4, PLP; 5, PL; 6, PN; 7, PA. Peak A is smaller than in Fig. 11.10, but still clearly seen, but peak B is no longer seen.

metabolites PL, PN, 4-PA and, to a much lesser extent, PLP, two other additional peaks named A and B were seen. They appear about 1 h after taking the pyridoxine and are then rapidly metabolised after about 2 h. We believe these two peaks to be new metabolites of pyridoxine which have previously been unrecognised. Figure 11.11 shows the plasma-B6 profile from a normal volunteer also on 800 mg PN/day for one week. This blood sample was taken about 4 h after the last vitamin supplement. Again, the unknown B6 metabolite (peak A) appeared with peak B absent on this occasion, probably due to its complete conversion to other metabolic forms. These two new B6 metabolites have been shown consistently to occur in the plasma of normal and hyperoxaluric individuals who were on high doses of pyridoxine (400 mg/day or greater).

We have shown in a separate HPLC experiment that the pyridoxine tablets used in this study are about 99% pure and it is highly unlikely that the additional peaks we were seeing earlier are impurities present in the tablets.

At the present time we do not know the chemical composition of these new metabolites, but a glance at Fig. 11.1 permits the following suggestion. It is seen that the pyridoxine molecule contains two hydroxy-methyl groups in positions 4 and 5. Apparently only the 5-hydroxy group is phosphorylated and only the 4-hydroxy group is aminated. It occurs to us that these chemical transitions might be reversed so giving rise to isomers of PMP, PLP, PL, PM, PMP and PA. Thus in theory there might be six new metabolites to be found rather than the two discovered so far.

Half-life of Plasma Pyridoxine

By plotting time on the horizontal axis and plasma pyridoxine on the vertical axis (log scale) a straight-line disappearance curve can be obtained from which the half-life of pyridoxine in plasma can be calculated. With normal volunteers taking single doses of 200–400 mg of pyridoxine, the half-lives prove to be 40 and 45 min respectively. When the half-life of pyridoxine was examined in the plasma of a patient with mild metabolic hyperoxaluria after a single oral dose of 200 mg, the half-life proved to be 40 min. This particular patient was already regularly taking 20 mg/day.

Effect of Dose of Pyridoxine on Plasma Levels of Vitamers

Each of a group of 9 volunteers was given increasing doses of pyridoxine by mouth. The doses were 100, 200, 400 and 800 mg per day. Each dose was given for 7 days after which the next dose was given for 7 days and so on.

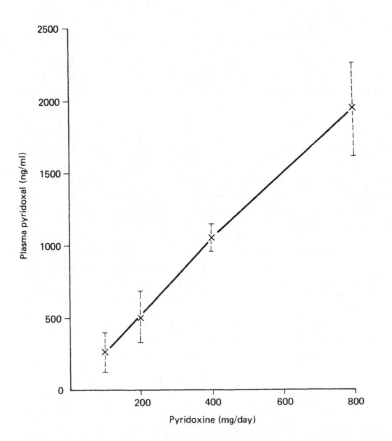

Fig. 11.12. Mean and SD of plasma PL levels after 7 days of doses of pyridoxine shown in 9 normal subjects (see text).

Venous blood samples were taken before the first dose and at the end of each dosage period. Plasma samples were analysed for vitamin-B6 metabolites by the HPLC method. The results for PL, PLP, and 4-PA are shown in Figs. 11.12–11.14. For 4-PA a straight-line relationship was found between pyridoxine given and the plasma level, and a virtually straight-line relationship was also found for PL. In the case of PLP, however, the situation was quite different. A very steep rise in plasma PLP was found for pyridoxine doses up to 100 mg/day, but this rapidly levelled off and in fact appeared to decline between

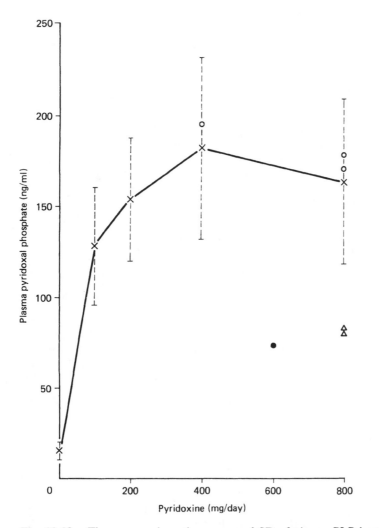

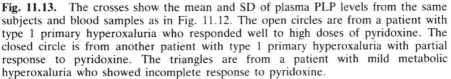

Fig. 11.13.   The crosses show the mean and SD of plasma PLP levels from the same subjects and blood samples as in Fig. 11.12. The open circles are from a patient with type 1 primary hyperoxaluria who responded well to high doses of pyridoxine. The closed circle is from another patient with type 1 primary hyperoxaluria with partial response to pyridoxine. The triangles are from a patient with mild metabolic hyperoxaluria who showed incomplete response to pyridoxine.

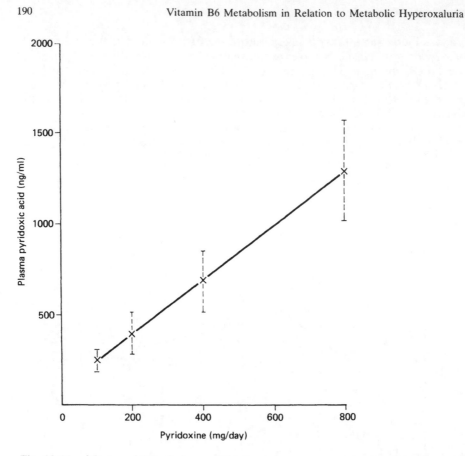

Fig. 11.14.   Mean and SD of plasma 4-PA from the same normal subjects as illustrated in Fig. 11.12.

dose levels of pyridoxine of 400 and 800 mg/day. However, in view of the large standard deviations it may be safer to say that the plasma PLP level reached a maximum of about 170 ng/ml at a dose of pyridoxine of about 300 mg/day.

Plasma Levels of Vitamin B6 in Metabolic Hyperoxalurias

This investigation of vitamin B6 metabolism was really initiated in order to see whether or not it is disturbed in either primary hyperoxaluria or mild metabolic hyperoxaluria. Although results so far are very few they are presented here. However, they should be regarded at the moment as preliminary. Fig. 11.13 shows 3 plasma levels from a patient with primary hyperoxaluria who had responded well to high doses of pyridoxine. It is seen that the points all lie very close to the curve for normal subjects. On the other hand, low levels of plasma PLP relative to dose of pyridoxine were found in a patient with type 1 primary hyperoxaluria and in two samples from a patient with mild metabolic hyperoxaluria. Both of these patients showed only partial response to pyridoxine and it is tempting to speculate that this was due to failure of plasma PLP to reach a sufficiently high level. However, this conclusion must wait upon further work which is in progress.

Urinary Excretion of Pyridoxic Acid after Load of Pyridoxine

Five normal individuals were each given 800 mg of pyridoxine by mouth for 5 days and the urine collected in the last 24 h. 4-PA levels were measured by the fluorimetric method and the 24-h excretions expressed as the percentage of the expected value if all the dose was excreted as 4-PA. Values were 40.3, 38.9, 32, 43, and 55.3% respectively, mean 41.9%. This value is in agreement with that found previously in a group of patients with hyperoxaluria (Rose and Samuell 1987).

References

Adams E (1979) Fluorimetric determination of pyridoxal phosphate in enzymes. Methods in Enzymology, vol 62. Academic Press, pp 404–410
Camp VM, Chipponi J, Faray BA (1983) Radioenzymatic assay for direct measurement of plasma pyridoxal-5′-phosphate. Clin Chem 29: 642–644
Coburn SP, Mahuren JD (1983) A versatile cation-exchange procedure for measuring the seven major forms of vitamin B6 in biological samples. Anal Biochem 129: 310–317
Dakshinamurti K, Chauhan MS (1981) Chemical analysis of pyridoxine vitamers. In: Leklem JE, Reynolds RD (eds) Methods in vitamin B6 nutrition. Plenum Press, New York, pp 99–122
Faber SR, Feitler WW, Bleiler AA, Ohlson MA, Hodges RE (1963) The effects of an induced pyridoxine and pantothenic acid deficiency on secretions of oxalic and xanthurenic acids in the urine. Am J Clin Nutr 12: 406–412
Gershoff SN (1964) The formation of urinary stones. Metab Clin Exp 13: 875–881
Gershoff SN, Faragalla FK, Nelson DA, Andrus SB (1959) Vitamin B6 deficiency and oxalate nephrocalcinosis in the cat. Am J Med 27: 72–80
Gibbs DA, Watts RWE (1970) The action of pyridoxine in primary hyperoxaluria. Clin Sci 38: 277–286
Gregory JF, Kirk JR (1981) Determination of vitamin B6 compounds by semi-automated continuous flow and chromatographic methods. In: Leklem JE, Reynolds RD (eds) Methods in vitamin B6 nutrition. Plenum Press, New York pp 149–170
Gregory JF, Manley DB, Kirk JR (1981) Determination of vitamin B6 in animal tissues by reverse-phase high-performance liquid chromatography. J Agric Food Chem 29: 921–927
Hamfelt A (1967) Enzymatic determination of pyridoxal phosphate in plasma by decarboxylation of L-tyrosine-C(U) and a comparison with the tryptophan load test. Scand J Clin Lab Invest 20: 1–10
Haskell BE (1981) An improved colorimetric assay for pyridoxal phosphate using highly purified apotryptophanase. In: Leklem JE, Reynolds RD (eds) Methods in vitamin B6 nutrition. Plenum Press, New York pp 69–78
Haskell BE, Snell EE (1972) An improved apotryptophanase assay for pyridoxal phosphate. Analyt Biochem 42: 567–576
Liu PKS, Rose GA (1988) A simple HPLC method for measuring plasma vitamin B6 compounds. Clin Chem In press
Lumeng L, Lui A, Li T-K (1981) Microassay of pyridoxal phosphate using tyrosine apodecarboxyl-ase. In: Leklem JE, Reynolds RD (eds) Methods in vitamin B6 nutrition. Plenum Press, New York pp 57–67

Lumeng L, Lui A, Oei TO (1984) Comparison of the use of L-tyrosine apodecarboxylase and D-serine apodehydratase for plasma pyridoxal phosphate assay. J Nutr 114: 385–392

Miller LT, Edwards M (1981) Microbiological assay of vitamin B6 in blood and urine. In: Leklem JE, Reynolds RD (eds) Methods in vitamin B6 nutrition. Plenum Press, New York pp 45–55

Morita E, Mizuno N (1980) Separation of vitamin B6 by reversed-phase ion-pair high-performance liquid chromatography. J Chromatogr 202: 134–138

Polansky M (1981) Microbiological assay of vitamin B6 in foods. In: Leklem JE, Reynolds RD (eds) Methods in vitamin B6 nutrition. Plenum Press, New York pp 21–44

Ribaya JD, Gershoff SN (1979) Inter-relationships in rats among vitamin B6, glycine and hydroxyproline. Effects of oxalate, glyoxylate, glycolate and glycine on liver enzymes. J Nutr 109: 171–183

Rose GA (1988) The role of pyridoxine in stone prevention of hyperoxaluric patients. In: Rous SN (ed) Urology Annual Part 2. Appleton and Lange, USA

Rose GA, Samuell CT (1987) The hyperoxaluric states. In: Rous SN (ed) Stone disease: diagnosis and management. Grune and Stratton, Orlando pp 177–205

Shin YS, Rasshofer R, Friedrich B, Endres W (1983) Pyridoxal-5'-phosphate determination by a sensitive micromethod in human blood, urine and tissues; its relation to cystathioniuria in neuroblastoma and biliary atresia. Clin Chim Acta 127: 77–85

Singkamani R, Worthington DJ, Thurnham DI, Whitehead TP (1986) A direct assay for pyridoxal-5'-phosphate using pig heart apoaspartate transaminase. Ann Clin Biochem 23: 317–324

Suelter CH, Wang J, Snell EE (1975) Application of a spectrophotometric assay employing a chromogenic substrate for tryptophanase to the determination of pyridoxal and pyridoxamine-5'-phosphates. Anal Biochem 76: 221–232

Takanashi S, Tamura Z (1970) Preliminary studies for fluorimetric determination of pyridoxal and its 5'-phosphate. J Vitaminol 16: 129–131

Thanassi JW, Cidlowski JA (1980) A radioimmunoassay for phosphorylated forms of vitamin B6. J Immunol Methods 33: 261–266

Vanderslice JT, Maire CE, Beecher GR (1981) Extraction and quantitation of B6 vitamers from animal tissues and human plasma: A preliminary study. In: Leklem JE, Reynolds RD (eds) Methods in vitamin B6 nutrition. Plenum Press, New York pp 123–147

Woodring MJ, Fisher DH, Storvick CA (1964) A microprocedure for the determination of 4-pyridoxic acid in urine. Clin Chem 10: 479–489

Yang I-Y, Sawhney AK, Pitchlyn RC, Peer PM (1981) Simple assay for femtomoles of pyridoxal and pyridoxamine phosphates. In: Leklem JE, Reynolds RD (eds) Methods in vitamin B6 nutrition. Plenum Press, New York pp 79–98

Subject Index